MONOGRAPHS OF THE PHYSIOLOGICAL SOCIETY

*Editors: H. Davson, A. D. M. Greenfield*
*R. Whittam, G. S. Brindley*

Number 6 PHYSIOLOGY OF THE RETINA
AND VISUAL PATHWAY

MONOGRAPHS OF THE PHYSIOLOGICAL SOCIETY

Volumes marked * are now out of print.

# PHYSIOLOGY OF THE RETINA
# AND VISUAL PATHWAY

## G. S. BRINDLEY
### M.D., F.R.S.

*Fitzmary Professor of Physiology in the University of London
at the Institute of Psychiatry*

Second edition

LONDON
EDWARD ARNOLD (PUBLISHERS) LTD

First published 1960
Second edition 1970

SBN: 7131 4176 x

2 6 / Z
OPTOMETRY

*Printed in Great Britain by*
*The Camelot Press Ltd., London and Southampton*

# PREFACE

T HE second edition of this book, like the first, is about the special physiology of the retina and visual pathway of man and of vertebrates that resemble him. Invertebrates are not considered, and the comparative physiology of vertebrates is used only in trying to decide whether observations made on other species are likely to be valid for man. Aspects of biochemistry and cellular physiology that are general for the whole nervous system receive only passing mention. Within these limits I have tried to be as comprehensive as was possible within my chosen length; that is, I have read most of the published literature of the subject up to about September, 1969, and tried to ensure that most omissions are intentional, and based on the speciality or unrelatedness of the facts omitted. With my own work I have been more lenient than with the work of others. This is not solely from vanity; readers have a right to expect that a book of this kind will lead them to the whole of the author's own contribution to the subject, even if they find it idiosyncratic in its choice of other material.

*London, 1970*                                            G. S. B.

# CONTENTS

# THE PHOTOCHEMISTRY OF THE RETINA

## The photolabile pigments of the retina

An organ whose function is to detect light and signal its presence
and pattern to the brain must contain a pigment, since light cannot
influence matter unless it is absorbed. The only exception to this
rule, the mechanical force exerted on a mirror by light reflected
from it, is too small by many orders of magnitude to be of physiological significance. We shall therefore expect to find, in any cell
capable of transmitting a signal to other cells under the influence
of light, a pigment which absorbs light of that range of wavelengths which can stimulate the cell. There is good evidence, as
we shall see in Chapter 6, that the light-sensitive structures of the
vertebrate retina are the rods and cones; and in the rods of many
species and the cones of a few, appropriate pigments have been
found.

It was H. Müller who in 1851 first described the reddish colour
of the outer segments of the rods of many vertebrate eyes. The
observation was neglected, and was made again by Boll in 1876.
Boll noted that the colour disappeared rapidly from the retina
after it had been removed from the animal, and described the
sequence of changes between the initial red and the final absence
of colour. He found that if the animal had been exposed to bright
light before it was killed, no redness was to be seen in the retina.
The importance of these observations was quickly appreciated by
Kühne, who in a series of papers (Kühne, 1878*a*, *b*, 1882*a*, *b*;
Ewald & Kühne, 1878*a*, *b*, *c*; Ayres & Kühne, 1882) showed that
a purplish-red pigment which he called 'Sehpurpur' or 'Rhodopsin' could be extracted from the retinas of a wide range of vertebrates with a solution of bile salt, and that the colour, both of the

retinas and of the pigment in solution, was stable in the dark but rapidly bleached by light of those wavelengths that it absorbed. Kühne also made many experiments on the chemical properties of the pigment and its regeneration after bleaching both in the retina and in solution.

There can be little doubt that, for those rods that contain it, rhodopsin is the pigment by which light has to be absorbed to cause transmission of a signal to the neural layers of the retina. In man and in a number of other species possessing rhodopsin, the spectral sensitivity curve of the eye in the fully dark-adapted state has been compared with the absorption spectrum of rhodopsin, and a very close agreement found (see p. 161). In some freshwater fish, whose rods contain instead of rhodopsin a different pigment, porphyropsin, the spectral sensitivity of the dark-adapted eye has been found to agree with the absorption spectrum of porphyropsin (Grundfest, 1932; Granit, 1941).

Pigments different from rhodopsin, but resembling it in being rapidly bleached by light, have been extracted from the retinas of a number of vertebrates. Most of these pigments are known or presumed to be derived from the outer segments of rods. Wald (1937; see also Wald, Brown & Smith, 1955) extracted from the retinas of poultry (which have more cones than rods) a photolabile substance that is very probably a cone pigment, and Bridges (1962a) has obtained a similar one from pigeons. Attempts to obtain solutions of indubitable cone pigments by using retinas that entirely lack rods (some diurnal reptiles) have so far been unsuccessful; but cone pigments can be detected by spectrophotometry of whole retinas or of isolated cones (see pp. 40–46).

There is no necessary reason why the pigment which allows a rod or cone to be influenced by light should be detectably decomposed by light. It could resemble a sensitizer in a photographic emulsion; the absorbing molecule passes the energy of an absorbed quantum immediately to a molecule of some other substance, and itself returns to the state in which it was before the quantum was absorbed, so that by ordinary criteria it is photostable. No photostable pigment likely to be concerned in the detection of light has yet been extracted from any retina, though Naka & Rushton (1966a) have argued that one is probably present in certain red-sensitive cones of the goldfish.

The great majority of the investigations that have been made of the chemical properties of visual pigments have used rhodopsin extracted from the retinas of frogs or cattle. The rhodopsins of frogs and cattle differ slightly in their absorption spectra, the maxima of which lie at 505 and 497 nm respectively; but in their other properties they, and the rhodopsins of other species, appear to be closely similar. In the two sections which follow, rhodopsin will be taken as a typical receptive pigment, and its properties in solution and within the rod will be described in detail.

## The properties of rhodopsin in solution

Rhodopsin is ordinarily extracted from a suspension of the outer limbs of rods, separated from other retinal tissue by centrifugation. The extracting agent is always an aqueous solution of a detergent, that is a substance whose molecules or ions are hydrophilic at one end and hydrophobic at the other. The most extensively used detergent is digitonin, a glycoside which has the advantage of being colourless; but there are many other satisfactory agents (Bridges, 1957; Crescitelli, 1967). In the absence of a detergent, rhodopsin is almost certainly insoluble in water, for it is precipitated from digitonin extracts if the digitonin is removed from them by dialysis (Broda, Goodeve & Lythgoe, 1940).

### Chemical nature of rhodopsin

Wald (1938) bleached solutions of rhodopsin with light and extracted them with acetone and light petroleum. In the extract he found a pale yellow carotenoid which he called *retinene*. Similar extraction of unbleached solutions of rhodopsin yielded no 'retinene'. Subsequently Morton & Goodwin (1944) proved that 'retinene' is the aldehyde of vitamin A. Vitamin A is now often called retinol, and 'retinene' must be called either retinal or retinaldehyde, since the termination -ene is reserved for unsaturated hydrocarbons. Wald's observations provide the simplest of the several pieces of evidence which together prove that one part of the molecule of rhodopsin is very closely related to retinaldehyde. The stereoisomeric configuration of this carotenoid part of the molecule and the way in which it is joined to the non-carotenoid part will be considered later.

It has long been generally supposed that the non-carotenoid

part of rhodopsin (called *opsin*) is a protein. Rhodopsin has a high molecular weight (see below), and is thermolabile, the kinetics of its thermal decomposition being similar to those of proteins (Lythgoe & Quilliam, 1938a; Hubbard, 1958). Solutions of it are bleached by proteolytic enzymes (Ayres, 1882; Radding & Wald, 1958), and precipitates of it, made by removing by dialysis the substances of low molecular weight from digitonin extracts of frog's rods, contain about 12 per cent of nitrogen, an amount not very different from that characteristic of proteins (Broda, Goodeve & Lythgoe, 1940). Its ultraviolet absorption spectrum shows a strong band around 280 nm, similar to that found in proteins which contain tyrosine and tryptophan (Collins, Love & Morton, 1952a). Hubbard (1969) used this spectroscopic property to estimate the amount of tyrosine and tryptophan present; the estimates agree with some (Bownds, 1967b) but not all (Heller, 1968; Shichi *et al.*, 1969) that have been obtained by analysing hydrolysed rhodopsin derivatives. Fifteen other amino-acids have been detected and estimated by Bownds, by Heller and by Shichi in hydrolysed rhodopsin derivatives; there are some quantitative discrepancies, but these will doubtless soon be resolved. The evidence leaves no doubt that the non-carotenoid part of the rhodopsin molecule is protein, but shows that besides protein and retinaldehyde it probably contains glucosamine, mannose and galactose. It may perhaps contain phospholipid (Krinsky, 1958a; Poincelot & Zull, 1969), but the analyses by Heller (1968), of chromatographically purified solutions, probably suffice to show that the phospholipid found in solutions purified by other methods is not chemically combined with the rhodopsin. We shall see later that after retinaldehyde has been removed from rhodopsin, the opsin that remains can combine again with retinaldehyde, provided that the latter is in the correct stereoisomeric form, to re-form rhodopsin.

### Molecular weight

Aqueous solutions containing rhodopsin and digitonin sediment in the ultracentrifuge as if they contained a homogeneous population of particles of molecular weight about 270,000 (Hecht & Pickels, 1938). All the rhodopsin and most of the digitonin sediments with these particles. They are thus presumably micelles containing molecules of both substances. Hubbard (1954) measured the extinction coefficient, at wavelength 500 nm, of solutions containing a known weight (and therefore a known number) of micelles. By using the data of Wald & Brown (1953) for the extinction coefficient of rhodopsin at this wavelength per mole of retinaldehyde liberated on bleaching, she was able to infer that each micelle contains only one retinaldehyde group. There must therefore be only one rhodopsin molecule in each micelle and one retinaldehyde group in each molecule. Hubbard also measured the nitrogen content of her solutions. Digitonin contains no nitrogen, and the amount of nitrogen in the solutions was such that, if it were all contributed by a typical protein containing 15% of nitrogen, the molecular weight of protein in each micelle would be 40,000. More recent estimates from amino-acid analyses of rhodopsin solutions purified chromatographically (Heller, 1968, 1969; Shichi et al., 1969) are between 27,000 and 30,000 for the rhodopsins of frogs, cattle and rats.

## Circular dichroism

Though retinaldehyde is a symmetrical molecule, opsin, like all proteins, is asymmetrical. In rhodopsin, the asymmetry shows itself by circular dichroism, i.e. a difference between the extent to which it absorbs left-handed and right-handed circularly polarized light (Crescitelli & Shaw, 1964; Crescitelli, Mommaerts & Shaw, 1966). The circular dichroism at wavelengths above 300 nm is positive (i.e. left circularly polarized light is absorbed more strongly than right), and is lost on bleaching. It is greatest at wavelengths near those where the total absorption is greatest, but wavelengths of maximal absorption and maximal circular dichroism do not coincide exactly, that for circular dichroism being about 15 nm less.

Rhodopsin has negative circular dichroism in the far ultraviolet between 210 and 230 nm, as do proteins known to have α-helical conformation. This circular dichroism is not lost on bleaching, but is diminished by 10 to 20%. Presumably the diminution represents a small change in the conformation of the protein part of the opsin. Besides this small change in circular dichroism, and a similar one in optical rotatory dispersion in the far ultraviolet, bleaching causes changes in the absorption spectrum in the same range (Takagi, 1963; Hubbard, Bownds & Yoshizawa, 1965).

## The effect of light on solutions of rhodopsin

When a molecule of rhodopsin absorbs a quantum of light, it is not split immediately to retinal and opsin, but passes, without requiring further action of light, through a long sequence of stages distinguishable by their characteristic absorption spectra. The order in which the various stages were discovered is not the order of their occurrence; more nearly it is the reverse. Different techniques, especially different temperatures, are appropriate for studying different stages, and the naming of some stages and their status as real members of the sequence is not free from controversy. The evidence for them will be presented in roughly historical order; but to make the reader's task easier a tentative conclusion from this evidence will be given first (Fig. 1.1).

*Indicator yellow and transient orange.* If a solution of frog rhodopsin, buffered to pH 9.3, is exposed to bright white light at room temperature for a few minutes, its purplish-red colour is found to be bleached to a very pale yellow. If the solution is then made acid, the colour alters reversibly to a deep yellow; a substance which acts as a pH indicator is present. This substance was named 'indicator yellow' by Lythgoe (1937). In alkaline solution the colour is stable for several hours, and acidification at the end of several hours produces the same depth of yellow as similar acidification immediately at the end of the bleaching period.

If, instead of bleaching for several minutes, the solution is exposed to a very bright light for a few seconds only and examined immediately, evidence is obtained for the existence of an intermediate stage in the conversion of rhodopsin into indicator yellow. When first examined after exposure to light the solution is orange. During the next few minutes, whether the solution is illuminated or not, the colour changes to yellow. The speed of change from orange to yellow decreases if the solution is cooled. Lythgoe & Quilliam (1938b) made use of this property in studying the unstable orange product of bleaching. They cooled solutions of rhodopsin to 3 °C, bleached them for three seconds with an extremely bright light, and then measured the absorption spectrum repeatedly. The results are shown in Fig. 1.2. Immediately after bleaching, the solution absorbed maximally at about 465 nm. During the follow-

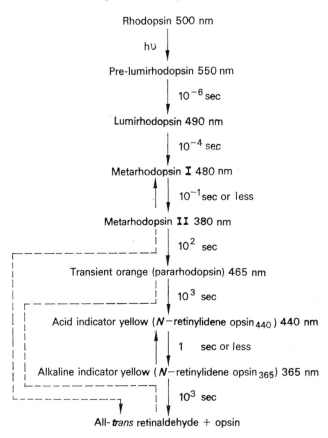

Rhodopsin 500 nm

hυ

Pre-lumirhodopsin 550 nm

$10^{-6}$ sec

Lumirhodopsin 490 nm

$10^{-4}$ sec

Metarhodopsin I 480 nm

$10^{-1}$ sec or less

Metarhodopsin II 380 nm

$10^2$ sec

Transient orange (pararhodopsin) 465 nm

$10^3$ sec

Acid indicator yellow ($N$-retinylidene opsin$_{440}$) 440 nm

1 sec or less

Alkaline indicator yellow ($N$-retinylidene opsin$_{365}$) 365 nm

$10^3$ sec

All-*trans* retinaldehyde + opsin

FIG. 1.1. Tentative scheme of the course of conversion of mammalian rhodopsin to all-*trans* retinaldehyde and opsin. After the name of each stage a rough figure is given for the maximum of its absorption band; the exact figure would vary with species and with temperature. Very rough estimates are given for the half-times of the transitions at 37°C. For the earlier transitions these are extrapolations from measurements made at much lower temperatures. The rough estimate of $10^{-1}$ sec for the metarhodopsin I ⟶ metarhodopsin II transition is for cattle rhodopsin in solution. An estimate for rabbit rhodopsin *in situ* would be two or three orders of magnitude smaller. Dotted arrows show unproven alternative sequences. Non-mammalian rhodopsins pass through corresponding stages, but the speeds of the transitions and the absorption maxima differ.

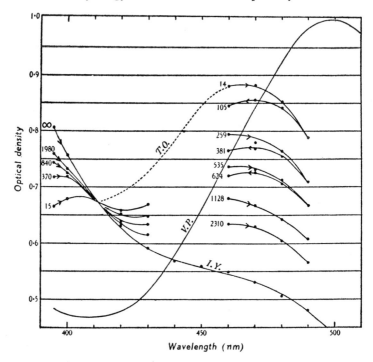

Fɪɢ. ɪ.2. The absorption spectrum of a solution of frog rhodopsin before bleaching (V.P., i.e. visual purple, a synonym for rhodopsin) and at intervals of from 14 to 2500 seconds after bleaching. pH 7·13, 3°C. The numbers against the points give the times (in seconds) at which the readings were taken; the arrows show the order in which they were taken. The dotted curve marked T.O. shows the estimated absorption spectrum of undecomposed transient orange, and that marked I.Y. the absorption spectrum after its thermal decomposition to indicator yellow is complete (from Lythgoe & Quilliam, 1938*b*).

ing 40 minutes, the absorption at wavelengths greater than 412 nm decreased, and that at wavelengths less than 412 nm increased. At 412 nm, it remained unchanged throughout the whole period of the experiment. A wavelength at which the optical density remains unchanged throughout the whole course of a process during which the densities at neighbouring wavelengths change is called an isosbestic point. Such a point should exist if the process is a single chemical reaction, provided that the sum of the optical densities due to the reactants is at some wavelengths greater than the sum of those due to the products, and at other wavelengths less, passing through a point of equality at an intermediate wavelength. No isosbestic point should be found if the process consists of a series of chemical reactions, unless all but one of them are extremely rapid, so that only the final products and the constituents of the stage preceding the one slow reaction are ever present in significant concentration. On the evidence of the well-defined isosbestic point in their experiments, Lythgoe and Quilliam argued that they were probably observing the conversion of a single substance, which they called transient orange, into indicator yellow.

*Lumirhodopsin and metarhodopsin.* Broda & Goodeve (1941) discovered that if a digitonin solution of rhodopsin is diluted with three times its volume of glycerol, cooled to $-73°C$, and then exposed to strong light, its absorption spectrum, which has been slightly and reversibly altered by the cooling, is further and irreversibly altered by the light. The changed spectrum is perfectly stable as long as the temperature is kept at $-73°$.

Broda and Goodeve supposed that the substance which remained stable at $-73°$ was the same as the transient orange of Lythgoe and Quilliam. Wald, Durell & St George (1950) reinvestigated it, and showed that it is in fact an earlier stage in the sequence. They named it 'lumirhodopsin', and showed that if it was warmed in darkness to about $-15°C$ it underwent a further change (see Fig. 1.3). The pigment then present, called 'metarhodopsin' by Wald, Durell and St George, is relatively stable at temperatures below $-15°C$, but if warmed in darkness to room temperature is converted to a mixture of approximately equal quantities of indicator yellow (or retinal and opsin) and a photolabile pigment closely resembling rhodopsin but not identical with it, to which Collins & Morton (who discovered it independently in the same year: 1950b) gave the name 'isorhodopsin'. We shall see later that the substances originally called 'metarhodopsin' and 'isorhodopsin' have been shown, chiefly by the work of the Harvard school (Wald, Hubbard and their associates), to be mixtures. The Harvard school have re-assigned the names to important single constituents of these mixtures.

Wald, Durell and St George found that a sequence of changes

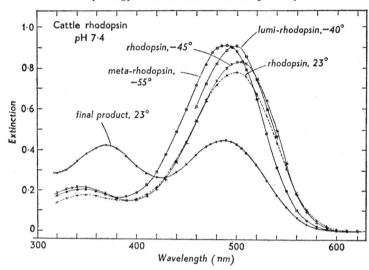

FIG. 1.3. Bleaching of cattle rhodopsin at low temperature in a buffered glycerol-water-digitonin solution. Absorption spectra are shown for the solution before exposure to light at 23° and −45°, after exposure to light at −40° (lumirhodopsin), after subsequent warming to −15° and re-cooling to −55° (meta-rhodopsin), and after warming to 23° (final product, consisting of a mixture of indicator yellow with a photolabile substance which Collins and Morton, who independently discovered these phenomena in the same year, called 'isorhodopsin', and which is almost certainly itself a mixture of rhodopsin and 9-*cis* rhodopsin). After the initial exposure to light, the solution was kept in darkness except for the very weak lights needed for measuring the absorption spectra (from Wald, Durell & St George, 1950).

very similar to those seen at low temperatures occurs at room temperature in dry films of rhodopsin in gelatin. Illumination produces a change in the absorption spectrum of these films which seems to correspond to the conversion of rhodopsin to lumi-rhodopsin. During the next 15 minutes the lumirhodopsin changes spontaneously into metarhodopsin, which is stable and unaffected by light in the dry state, but on moistening is converted to a photolabile mixture which probably contains rhodopsin, iso-rhodopsin and indicator yellow.

*Metarhodopsin II.* Matthews, Hubbard, Brown & Wald (1963) showed that metarhodopsin is a pH indicator whose shift of absorption maximum with pH is opposite in direction to that of Lythgoe's indicator yellow. Figure 1.4 shows the effect of pH on the absorption spectrum of metarhodopsin at 3·2°C. Increase of temperature favours the acid form ('metarhodopsin II'), so that the proportion of the acid form in the equilibrium mixture at pH 7, which is about 40% at 3·2°C, becomes about 70% at 20°C.

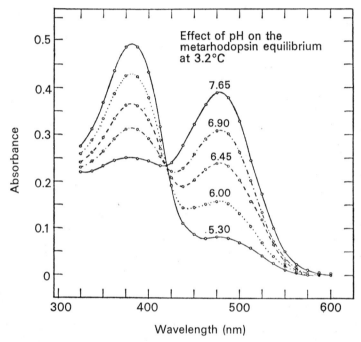

FIG. 1.4. The equilibrium between metarhodopsin I and metarhodopsin II (from cattle) at 3·2°C. At pH 7·65 most of the metarhodopsin is in the I form ($\lambda_{max}$ 480 nm); at pH 5·30 most is in the II form ($\lambda_{max}$ 3·80 nm) (from Matthews, Hubbard, Brown & Wald, 1963).

It seems that when rhodopsin is bleached by light the alkaline form ('metarhodopsin I') is always formed first, and this passes quickly into the equilibrium mixture appropriate to the pH and temperature. The further thermal changes that ultimately yield retinal and opsin are a good deal slower.

*Pre-lumirhodopsin.* If rhodopsin is exposed to light at temperatures even lower than those used by Broda and Goodeve and by

Wald, Durell and St George, an intermediate between rhodopsin
and lumirhodopsin with maximum absorption at about 550 nm
can be detected (Grellman, Livingston & Pratt, 1962; Yoshizawa
& Wald, 1963). It is called prelumirhodopsin, and is stable below
about −190°C.

### The photosensitivity of rhodopsin in solution

Schneider, Goodeve & Lythgoe (1939) measured at a number
of wavelengths the rate at which solutions of rhodopsin were
bleached by light of known intensity. In these circumstances it
is possible, from measurements of the optical density of the solutions without knowledge of their molar concentration, to calculate
the product of $\gamma$, the quantum efficiency of bleaching (i.e. the
number of molecules bleached for each quantum absorbed) and
$\epsilon$, the molar extinction coefficient (i.e. the logarithm to base 10 of
the ratio of the light intensities before and after passing through
one centimetre of solution, divided by the molar concentration of
the solution).

Figure 1.5 shows the values of $\epsilon\gamma$ obtained by Schneider,
Goodeve and Lythgoe at a number of wavelengths. It also shows,
on the same axes, the absorption spectrum of a solution of
rhodopsin, the vertical scale being adjusted so that the curve
passes through the point representing the photosensitivity at
502 nm. The close agreement between the points and the curve
over the range 436 to 560 nm shows that over this range the
quantum efficiency $\gamma$ is constant: any quantum that is absorbed,
whether it is of yellow light with energy $h\nu = 3\cdot5 \times 10^{-12}$ ergs or
of violet light with energy $h\nu = 4\cdot5 \times 10^{-12}$ ergs, has the same
probability of bleaching a molecule of rhodopsin. At shorter
wavelengths, the quantum efficiency becomes much smaller. This
is doubtless because at short wavelengths most of the quanta are
absorbed not by the carotenoid part of the molecule ('chromophore') but by the opsin. Such absorptions are less likely to lead
to bleaching, but probably not vastly less likely, for Kropf (1967)
found that the quantum efficiency at 280 nm was about one-third
of that in the visible part of the spectrum. The retinaldehyde
liberated by such bleaching with ultraviolet light is in the all-*trans*
configuration and the opsin is undamaged, as judged by its ability
to combine with 11-*cis* retinal to form rhodopsin. From Kropf's
experiments, either the chromophore can absorb substantial

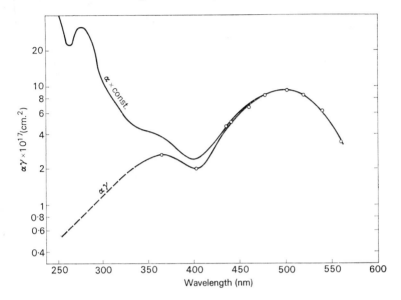

FIG. 1.5. The photosensitivity, $\alpha\gamma$ of frog-rhodopsin, compared with its absorption spectrum. $\alpha$ is the Napierian molecular extinction coefficient. To convert to the decadic molar extinction coefficient $\epsilon$, multiply by $1/1000 \times$ Avogadro's number $\times \log_e$, i.e. by $2\cdot6 \times 10^{20}$ (from Goodeve, Lythgoe & Schneider, 1941).

amounts of light of wavelength 280 nm (but this is unlikely; see Hubbard, 1969) or, as Kropf concludes, energy can be transferred rather efficiently from the opsin to the chromophore and cause it to undergo *cis-trans* isomerization.

When Schneider, Goodeve and Lythgoe first measured the photosensitivity, no information was available on the molar extinction coefficient of rhodopsin. Subsequently, Wald & Brown (1953) measured the extinction coefficient per mole of retinaldehyde liberated and found it to be 40,600. On the evidence of Hubbard (1954) there is one labile retinaldehyde group in each molecule of rhodopsin, so that the extinction coefficient per mole of retinaldehyde must be the same as the molar extinction coefficient. The measurements by Wald and Brown of $\epsilon$, taken together with those by Schneider, Goodeve and Lythgoe of $\epsilon\gamma$, lead to the estimate $0\cdot58$ for the quantum efficiency of bleaching in the range of the spectrum over which it is constant. Kropf (1967), from his own measurements of photosensitivity, finds a quantum efficiency of 0·66 at 436 nm. These estimates are not within experimental error of 1, and the bleaching was done so slowly that the difference cannot be explained by photic reconversion of metarhodopsin to rhodopsin (see p. 22); it is clear that when a molecule of rhodopsin has absorbed a quantum of light and thereby been raised into an excited state, there is a significant probability, which is almost independent of the energy of the quantum provided that it is low enough for absorption to be by the chromophore and not the protein, that it will not pass along the sequence of thermal reactions that leads to bleaching, but will revert spontaneously to the ground state.

### The production of retinol (vitamin A) in bleached extracts

In Lythgoe's experiments on digitonin extracts of washed rods, indicator yellow was the final product of bleaching. As long as the solution was kept alkaline, it was apparently stable, and if it was made acid, it faded slowly to a very pale yellow colour which we may presume was due to free retinaldehyde.

If, however, a fresh digitonin extract of whole retinas is bleached at pH between 5·5 and 8, the solution slowly becomes colourless. Extraction of this colourless solution with methanol and light petroleum yields only a trace of retinaldehyde, but a large amount of retinol. The formation of retinol from rhodopsin after bleaching was first demonstrated in solution by Bliss (1948), but had earlier been detected in whole retinas by Wald (1935; see p. 31). Freshly bleached extracts yield little or no retinol on extraction with methanol and light petroleum; the retinol is formed slowly in darkness after bleaching, presumably by reduction of retinaldehyde. Enzyme systems that perform this reduction are evidently present in extracts of whole retinas, and absent or nearly so in extracts of washed rods.

### The regeneration of rhodopsin in solution

When a thoroughly bleached solution of rhodopsin is allowed to stand in the dark at room temperature for several hours, some regeneration of rhodopsin may occur. The amount regenerated is never a large fraction of the amount present before bleaching, and sometimes none is regenerated. The factors determining whether regeneration occurs, and causing it never to exceed about 15%, remained obscure until a series of investigations by Hubbard and Wald, which have made possible a clear understanding of the conditions for the efficient synthesis of rhodopsin by systems of well-characterized constituents in aqueous solution. The knowledge gained is almost certainly closely relevant to the regeneration of rhodopsin in the living retina.

Hubbard & Wald (1951) found that a system containing purified opsin from the retinas of cattle, crystalline alcohol dehydrogenase derived from horse liver, retinol from fish liver oil, and cozymase (nicotinamide-adenine dinucleotide), brought together in aqueous digitonin, synthesized a satisfactory amount of rhodopsin. Alcohol dehydrogenase was already known to catalyse the reversible oxidation of retinol to retinaldehyde (Bliss, 1949), and it seemed that in the presence of opsin this reaction was occurring, and the retinaldehyde produced was combining with the opsin to produce rhodopsin.

Later, Hubbard & Wald (1952) tried the same experiment using crystalline retinol instead of fish liver oil concentrate, and obtained almost no rhodopsin. Retinol can exist in a number of stereoisomeric forms, involving different *cis-trans* configurations at the double bonds of the side-chain (see Fig. 1.6). In the crystalline retinol used, all of these were known to be in the *trans* configuration; and it was known that, in the presence of a trace of iodine, light is an effective means of converting a pure carotenoid into a mixture of its geometrical isomers. The crystalline all-*trans* retinol was therefore illuminated in the presence of a trace of iodine. It

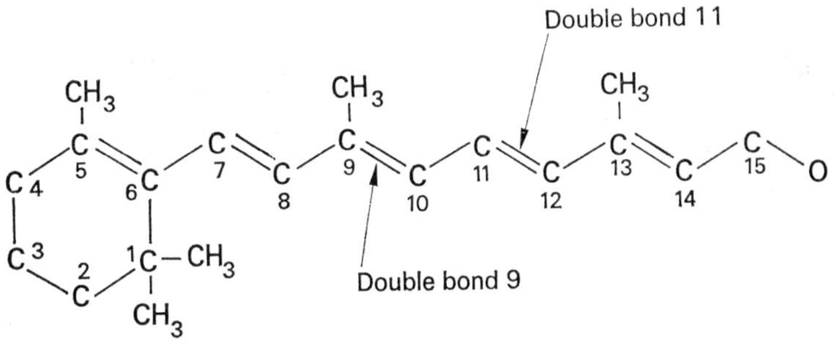

FIG. 1.6. The structure of retinaldehyde. The formula shows the all-*trans* configuration. The arrows point to the two double bonds (9 and 11) at which *cis-trans* isomerization is important in the retina.

then gave as good a yield of rhodopsin as the fish liver oil concentrate. Alcohol dehydrogenase was found to be effective in oxidizing both all-*trans* and liver oil retinol; thus the isomeric specificity in rhodopsin synthesis could not lie in the oxidation stage.

The coupling of retinaldehyde with opsin was next investigated. All-*trans* retinaldehyde, if carefully screened from light, was found to give only a trace of rhodopsin when mixed with opsin; but if the retinaldehyde was first illuminated, a good yield was obtained. Illuminated all-*trans* retinaldehyde was fractionated chromatographically, and a crystalline substance isolated ('neoretinene *b*', i.e. 11-*cis* retinaldehyde), which when mixed with excess of opsin was almost completely converted to rhodopsin. Two other isomers of retinaldehyde, 13-*cis* and 9, 13 *dicis*, were found not to react with opsin. A fifth isomer, 9-*cis*, was found to combine with opsin to give a photolabile pigment different from rhodopsin, which Hubbard and Wald called 'isorhodopsin'. They suggested that the substance obtained on allowing metarhodopsin to stand at room temperature, to which Collins and Morton had given the name 'isorhodopsin' two years before, was probably a mixture, of which the compound of 9-*cis* retinaldehyde with opsin was a principal constituent. Hubbard and Wald's re-assigning of the name 'isorhodopsin' is now generally accepted.

When solutions of rhodopsin are bleached by light of wavelength too long to be absorbed by retinaldehyde and isomerize it, the retinaldehyde that is liberated is found to be in the all-*trans* configuration. Figure 1.7 shows the combination of 11-*cis* retinaldehyde and opsin to form rhodopsin, and the decomposition of the synthesized rhodopsin by light into opsin and all-*trans* retinaldehyde. 11-*cis* and all-*trans* retinaldehyde differ in their absorption spectra, the most striking difference being the larger maximum specific extinction of all-*trans* retinaldehyde. In synthesis, the curves of Fig. 1.7 show an isosbestic point, suggesting that the combination of 11-*cis* retinaldehyde and opsin yields rhodopsin in one stage, or at least without any intermediate compound ever being abundant enough to affect the absorption spectrum appreciably. In bleaching there is no isosbestic point, as was to be expected from the known occurrence of intermediate compounds.

The fact that it is all-*trans* retinaldehyde that is liberated by bleaching explains the failure of bleached solutions to regenerate

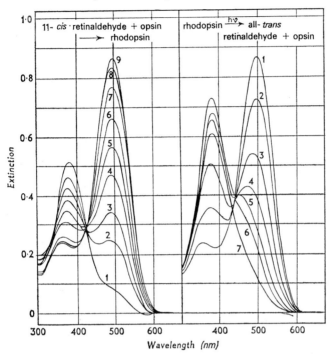

FIG. 1.7. The synthesis and bleaching of rhodopsin in solution. pH 7·0, 22·5°C. *Left:* a mixture of 11-*cis* retinaldehyde and cattle opsin was incubated in the dark, and absorption spectra recorded at intervals. Curve 1 was recorded at 0·3 min, and the others in order at 2·5, 5, 10, 18, 30, 60, 120 and 180 min. *Right:* the rhodopsin formed in the experiment shown on the left was exposed to light for various durations, and spectra recorded immediately after each exposure. Curve 1 shows the spectrum before illumination, and curves 2–6 in order after 5, 10, 15, 30 and 120 sec. Curve 7 was taken after further exposure to a stronger light (from Wald & Brown, 1956).

satisfactorily in the dark. It also explains the otherwise surprising finding of Chase & Smith (1939) that the amount regenerated is greater when the bleaching light is blue than when it is green or yellow; blue light isomerizes retinaldehyde.

*Retinaldehyde isomerase*

Ewald & Kühne (1878*b*) found that solutions of frog rhodopsin regenerate better after bleaching if they have been made from retinas and pigment layers together than if they have been made from retinas alone. Bliss (1951) showed that detergents are not necessary to extract the active substance from the pigment layers; saline extracts of pigment layers, if added to digitonin solutions of rhodopsin, greatly increase the extent to which they regenerate after bleaching. The activity is destroyed by boiling, is retained by collodion membranes in dialysis against 0·5% sodium chloride solution, and appears in the precipitate if the proteins of the extracts are precipitated with ammonium sulphate.

The significance of these observations was made clear by Hubbard (1956), who showed that saline extracts of the pigment epithelium of frogs or the retinas of cattle contain an enzyme which catalyses the conversion in darkness of all-*trans* or 11-*cis* retinaldehyde to a mixture of 95% all-*trans* and 5% 11-*cis*. The same extracts also increase about fivefold the rate of conversion of all-*trans* retinaldehyde into 11-*cis* by light. They have no action on the isomers of retinol or on the 9-*cis* or 13-*cis* isomers of retinaldehyde. Both the enhancement of photoisomerization and the catalysis of spontaneous isomerization in the dark of which Hubbard's retinaldehyde isomerase is capable are likely to have made their contribution to providing the 11-*cis* retinaldehyde necessary for regeneration in the experiments of Ewald and Kühne and of Bliss. For the regeneration of rhodopsin in the living eye, however, it seems that photo-isomerization of free retinaldehyde is of very little importance, and the useful function of retinaldehyde isomerase in the living eye is almost certainly to catalyse non-photic isomerization.

*The nature of the substances produced by the action of light on rhodopsin*

The substances named in the preceding pages as stages in the breakdown of rhodopsin to all-*trans* retinaldehyde and opsin are known primarily by their absorption spectra. They have not been purified, but a good deal of ingenuity has been used in obtaining indirect evidence about their structure. They will be considered here in reverse order, beginning with those that occur latest in the sequence of thermal reactions.

*Indicator yellow.* Wald (1938) found that from a solution of rhodopsin, bleached to the indicator yellow stage, extraction with acetone and light petroleum yielded a yellow compound which was later (Morton & Goodwin, 1944) shown to be retinaldehyde. Indicator yellow is not the same as retinaldehyde, for retinaldehyde is not a pH indicator; but Collins & Morton (1950a) found that, if egg albumin is added to it, indicator properties appear. This suggested the possibility that indicator yellow might not be a direct derivative or rhodopsin, but might result from the combination of free retinaldehyde with any protein (including opsin) that happens to be present in the solution. Whether this is so or not is of some importance in considering how retinaldehyde is linked to opsin in the rhodopsin molecule: if indicator yellow is a direct derivative of rhodopsin, any information about its structure is relevant to the problem of the structure of rhodopsin, but if it is not a direct derivative, its structure is of little physiological interest. The very close similarity in spectroscopic properties between indicator yellow and Schiff's bases produced by combination of retinaldehyde with primary amines (Ball, Collins, Dalvi & Morton, 1949; Morton & Pitt, 1955) makes it nearly certain that in molecules of indicator yellow the retinaldehyde is joined to protein by a CH : N. link in the alkaline form, and by a CH : N+H. link in the acid form.

The ready combination of retinaldehyde with $NH_2$ groups provides a possible mechanism by which retinaldehyde might appear as a short-lived intermediate compound in the formation of indicator yellow. Collins (1953) tested this possibility by illuminating solutions of rhodopsin at pH 9 in the presence of a large excess of formaldehyde, calculated to block nearly all the amino groups of the opsin. The absorption spectrum, measured soon after illuminating, was found to be intermediate between those of alkaline indicator yellow and retinaldehyde. On standing at 18°C, it altered in the direction of retinaldehyde, the change following unimolecular kinetics with a half-time of 9 minutes. The proportion of the total retinaldehyde which was in the form of indicator yellow at the first measurement was 40% or more, and extrapolation of the time-course of its spontaneous conversion to free retinaldehyde was consistent with the hypothesis that all the free retinaldehyde present was initially derived from indicator yellow. This experiment leaves little doubt that, at pH 9 in the presence of a large excess of formaldehyde, indicator yellow is formed from rhodopsin after exposure to light without passing through the intermediate stage of retinaldehyde. A similar argument can be applied to the formation of acid indicator yellow on exposure of rhodopsin to light in the absence of formaldehyde at pH less than 6. Most of the amino groups of the protein are then in the form $NH_3^+$, and these would not be expected to be capable of combining with the CHO groups of retinaldehyde.

If, as is likely, the only effect of the formaldehyde in Collins's experiment is to block amino groups, then the apparent stability of indicator yellow at pH 9 is not a true stability of the molecules originally formed from rhodopsin. These are dissociating into retinaldehyde and opsin with a half-time of 9 minutes at 18°C, and new indicator yellow is being formed at the same speed by combination of retinaldehyde with amino groups, which may not be the same as those to which it was originally attached. On this hypothesis, the instability of indicator yellow between pH 4 and pH 6 results from the lack of $NH_2$ groups with which the retinaldehyde liberated can recombine. In strongly acid solution (pH 3·3), indicator yellow is again stable (Lythgoe, 1937). Morton & Pitt (1955) suggest that molecules of the acid form of indicator yellow are always stable; manifest instability is found only when some of the molecules are in the alkaline form, and their environment does not include abundant free $NH_2$ groups.

On the basis of Collins's evidence that indicator yellow is formed from rhodopsin on exposure to light without the intermediary liberation of retinaldehyde, and of the still clearer evidence that a closely similar substance is formed, again without intermediary liberation of retinaldehyde, when rhodopsin is exposed to low pH in the absence of light, Morton & Pitt (1955) argued that in rhodopsin the aldehydic carbon atom of retinaldehyde is linked to a nitrogen atom in the opsin by a double bond, i.e. a Schiff-base linkage. Their conclusion has been generally accepted and is probably correct; but a possible ground for doubt has been introduced by the discovery (Poincelot et al. 1969, see p. 21) that the linkage of retinaldehyde to the ε-amino group of lysine demonstrable by hydrolysis in metarhodopsin II (and presumably in indicator yellow) is not demonstrable by the same technique in rhodopsin or in metarhodopsin I.

*Transient orange (pararhodopsin)*. The transient orange that Lythgoe and Quilliam identified as a product of bleaching frog rhodopsin is almost certainly the counterpart of the product of bleaching cattle rhodopsin that was called 'the 465 nm pigment' by Matthews *et al.* (1963) and 'metarhodopsin 465' by Ostroy *et al.* (1966). Wald (1968) has proposed the name 'pararhodopsin'. Frog transient orange is a good deal less stable than cattle pararhodopsin, but this is not at all incompatible with the supposition that they are corresponding stages in essentially similar sequences of chemical changes.

It is not certain whether pararhodopsin is a necessary stage in the breakdown of cattle rhodopsin to retinaldehyde and opsin. Ostroy *et al.* (1966) argue that it is, but Matthews *et al.* (1963) prefer the view that metarhodopsin II can be (and mostly is) converted to retinaldehyde and opsin without passing through a pararhodopsin stage. Though it is difficult to be sure whether pararhodopsin is an obligatory intermediate, there is no doubt (see p. 29) that in the intact retina a substance closely resembling it is sometimes formed in large quantities.

*Metarhodopsin II.* The pH-sensitive reversible interconversion of metarhodopsin I and metarhodopsin II has the property, at first sight surprising, that the shift of absorption band with pH is in the opposite direction to that found in synthetic Schiff bases of retinaldehyde. The property was made less surprising by the evidence of Poincelot, Millar, Kimbel & Abrahamson (1969) that metarhodopsin I and metarhodopsin II are not protonated and unprotonated forms of the same Schiff base: extraction of metarhodopsin II with methanol containing $10^{-3.5}$ hydrogen chloride yields *N*-retinyl lysine, as was to be expected from the earlier work of Bownds (1967*a*); but similar extraction of metarhodopsin I or rhodopsin yields *N*-retinyl phosphatidylethanolamine. Thus the site to which retinaldehyde is bound in rhodopsin and metarhodopsin I appears to be different from that to which it is bound in metarhodopsin II. There are, however, difficulties in accepting that it is bound to phosphatidylethanolamine in rhodopsin, since Heller (1968, 1969) could detect no phospholipid in highly purified rhodopsin solutions. It may be that the retinaldehyde group migrates during extraction with methanol and hydrogen chloride, or that rhodopsin contains phosphatidylethanolamine, but in quantity too small (less than 2%) for Heller to detect it.

B

*Metarhodopsin I.* The nature of metarhodopsin I and its relation to rhodopsin and to the isorhodopsin of Collins and Morton were clarified by Hubbard & Kropf (1958). They accepted the argument of Morton and Pitt that in all these compounds the retinaldehyde is attached to opsin through a Schiff-base linkage, and suggested that different stereoisomers form compounds differing in stability and spectroscopic properties according to the degree of fit of the retinaldehyde to its appropriate site in the opsin molecule. 11-*cis* and 9-*cis* retinaldehyde were already known to combine with opsin to produce the thermally stable but photolabile pigments rhodopsin and isorhodopsin (see p. 16). Hubbard and Kropf suggested that in these photolabile pigments the chromophore retains its 11-*cis* or 9-*cis* configuration, which fits well the site of attachment in the opsin molecule. The first action of light is to isomerize the chromophore to the all-*trans* configuration, in which it fits the opsin molecule badly and is liable, at temperatures exceeding about − 20°C, to become detached, yielding (ultimately) free all-*trans* retinaldehyde and opsin; but whilst the all-*trans* retinaldehyde is still attached to the opsin, the absorption of another quantum of light by it may cause reisomerization to another configuration, and if this is 11-*cis* or 9-*cis*, rhodopsin or isorhodopsin will be re-formed. Thus the result of exhaustively illuminating rhodopsin at temperatures below about − 20°C will be a mixture of stereoisomeric retinaldehyde-opsin compounds, their proportions depending on the photosensitivities for the several transitions. Since the absorption spectra of these compounds differ, the composition of the steady-state mixture obtained should depend on the wavelength of the light to which it has been exhaustively exposed. Hubbard and Kropf find that it does, and this is a part of the evidence in favour of their theory.

On the theory of Hubbard and Kropf, metarhodopsin as hitherto understood is thus a mixture of thermally stable rhodopsin and isorhodopsin with thermally unstable compounds of opsin with other retinaldehyde isomers, chiefly all-*trans*. They have reassigned the name metarhodopsin to the thermally unstable part of this mixture. The isorhodopsin of Collins and Morton is, on the Hubbard–Kropf theory, the thermally stable part of the steady-state mixture, and consists of roughly equal quantities of rhodopsin and isorhodopsin. This photolabile material, found after rhodopsin illuminated at $-30°C$ or less has been warmed to room temperature, is not, as Collins & Morton (1950*b*) and Wald, Durell & St George (1950) originally supposed, regenerated in the dark from the products of photolysis; the greater part of it is present all the time, and is merely unmasked by the thermal decomposition of the compounds of opsin with retinaldehyde isomers other than 11-*cis* and 9-*cis*. There is, however, also some regeneration under these conditions (Bridges, 1960).

*Lumirhodopsin and pre-lumirhodopsin.* The isomerization of the chromophore from 11-*cis* to all-*trans* has probably occurred by the pre-lumirhodopsin stage. If so, the thermal conversion of pre-lumirhodopsin through lumirhodopsin to metarhodopsin does not affect the steric configuration of the chromophore, but either that of the opsin or the relation of the chromophore to the opsin.

## The organization of rhodopsin in the rods

Within the rods, rhodopsin is present in very high concentration. From measurements of the amount that could be extracted with digitonin, Hubbard (1954) estimated that in cattle about 4% and in the frog about 10% of the wet weight of a rod outer segment consists of rhodopsin. From later literature, Blaurock & Wilkins (1969) derive the estimate that rhodopsin constitutes between a seventh and a quarter of the protein of the outer segment. The packing of rhodopsin in the rods is not random. Rods are dichroic, absorbing plane-polarized light much more strongly if its electric vector is perpendicular to their long axes than if it is parallel to them (Schmidt, 1938; Denton, 1954). The dichroism indicates that the chromophore groups of the molecules of rhodopsin must all or nearly all be orientated in a plane perpendicular to the long axes of the rods. Two questions remain to be considered elsewhere: whether the chromophores are free to rotate within this plane (Brindley, 1960*b*, p. 161) and whether all or most of the molecules face in the same direction, or half in one direction and half oppositely (p. 32).

Further information about the internal organization of rods has been obtained by electron-microscopical examination of thin sections (Sjöstrand, 1953; de Robertis, 1956; Dowling, 1965;

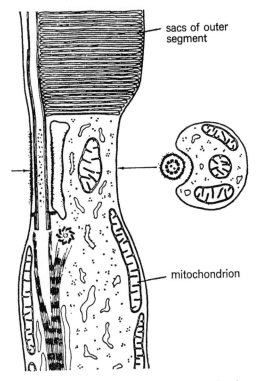

sacs of outer
segment

mitochondrion

FIG. 1.8. The fine structure of the junctional region between the
outer and inner segments of a rod, as inferred from electron
micrographs. The transverse section (small picture on the right)
shows how the outer segment (top of the main picture) is joined
to the inner segment by a filament with the characteristic internal
structure of a cilium, and also by a broader cytoplasmic bridge
(from Richardson, 1969).

Richardson, 1969) and of fragmented rod outer segments (Brown,
Gibbons & Wald, 1963; Pedler & Tilly, 1967). Figure 1.8 shows
an interpretation of such electron micrographs. The main part
of the rhodopsin-containing outer segment consists of a pile of
flattened sacs. In the guinea-pig (Sjöstrand), the thickness of each
membrane is 30 Å, the space between membranes within the sac
80 Å, and the space between one sac and the next 100 Å, so that
the pattern repeats itself every 240 Å. The length of the rod is
about 17 µm, so that there are about 700 sacs, or 1400 membranes.

Most of the sacs in rods appear to be closed, and their membrane not continuous at any point with the surface membrane of the rod or the membranes of other sacs; but it is difficult to be quite sure of this, and the hypothesis, attractive when one tries to understand how excitation may be conducted, that all sac membranes are continuous with one another, may not be firmly disproved. A few sacs near the basal parts of the outer segments of some rods, and a large proportion of the sacs of cones (which have the same general type of structure) are demonstrably continuous with the surface membrane. Experiments of Cohen (1968) on the penetration of lanthanum from extracellular space into sacs support the view that continuity with the surface membrane is usual in cone sacs and rare in rod sacs.

The manner in which the molecules of rhodopsin are related to the structures visible with the electron microscope is imperfectly known, but it is a reasonable hypothesis that each sac contains (and is to a large extent made up of) layers of regularly orientated rhodopsin molecules. A protein molecule of molecular weight 30,000 should, if spherical, have a diameter of about 45 Å. This would be compatible with as many as five monomolecular layers per sac in the guinea-pig and ten in the frog, but it is likely that the number is smaller. X-ray diffraction analysis has been applied to intact wet frog retina by Blaurock & Wilkins (1969). On the assumption (based on published electron micrographs) that each rod sac has a mirror plane, they obtain an electron density map which would correspond well to two protein layers and one hydrocarbon layer per sac membrane. Both protein layers or only one might contain rhodopsin; on the former supposition there would be four layers of rhodopsin per sac.

The refractive index of isolated rods and cones and of their component parts has been estimated by Sidman (1957) by suspending them in solutions of serum albumin of varying strength (and therefore of varying refractive index), and finding at what strength the rod or cone constituents are invisible by phase contrast microscopy. For the outer segments of rods in a number of species the refractive index is about 1·41, and for the outer segments of cones about 1·385. The extracellular fluid which separates outer segments in the living retina presumably has the same refractive index as physiological saline, i.e. 1·334.

## The bleaching and regeneration of rhodopsin in the rods

It ought not to be assumed without evidence that the physical properties of rhodopsin, the sequence of chemical changes that follow absorption of light by it, and the factors that determine its regeneration, are necessarily the same when it is organized in a quasicrystalline structure in the rod as when each molecule is isolated from its neighbours and surrounded by a shell of digitonin molecules in aqueous solution. The absorption spectrum may be shifted a few nanometres to the red; difference spectra for human rods and rhodopsin measured in the presence of hydroxylamine, which should convert all the coloured products of photolysis to colourless retinaldehyde oxime, show such a shift clearly (Wald & Brown, 1958). However, in the frog, carp and conger eel, Dartnall (1961) found no difference between the maxima of the direct absorption spectra of rods and of the pigments extracted from them under conditions where a shift as large as that found by Wald and Brown should have been easily detectable. It is difficult to be sure whether this is a species difference, or whether, as Dartnall suggests, the shift found by Wald and Brown is due, despite the hydroxylamine, to a difference in the products of photolysis in the two conditions.

The sequence of events that follows the absorption of light by rhodopsin in the rods is less well understood than its counterpart for aqueous solution, but the available evidence indicates that it is very similar. The immediate consequence of absorption of a quantum must almost certainly be the same. For the later changes we have evidence from spectrophotometry of suspensions of rods, whole isolated retinas, and intact living eyes, from the results of chemical extraction of illuminated isolated retinas (see p. 31), from the early receptor potential (see p. 31), and from techniques that permit very rapid measurement of the pH and conductance of packed suspensions of rods (see p. 33).

*Spectrophotometry of rod suspensions and isolated retinas.* Kühne (1878b) observed that the changes of colour which followed exposure to light were similar for whole frogs' retinas and for extracts from them made with solutions of bile salt. Spectrophotometric examination of isolated whole retinas confirms this similarity. Dark-adapted retinas of several vertebrates have been shown to have absorption spectra very similar to those of well-purified solutions of single visual pigments, provided that precautions are taken to collect on the photocell almost all the light that has passed through the rods (Denton & Walker, 1958). The sequence of events that follows bleaching has been studied by this technique (Brindley, 1960b; Frank & Dowling, 1968; Cone & Brown, 1969), by transmission spectrophotometry in whole excised albino eyes (Hagins, 1958; Cone & Cobbs, 1969), and in suspensions of rods (Arden, 1954; Bridges, 1962b). In the frog (Brindley, 1960b), the unbleached retina has an absorption spectrum closely resembling that of frog rhodopsin, the wavelength of maximum absorption being $503 \pm 2$ nm. A minute after thorough exposure to white light at 20°C, a strong absorption around 380–420 nm is found, such as might be due to meta-rhodopsin II and free retinaldehyde. There is also a subsidiary peak at about 480 nm, which probably corresponds to transient orange (pararhodopsin). If the retina is left in the dark, the absorption at 380–420 nm diminishes to about $\frac{1}{2}$ in 10 minutes and to about $\frac{1}{4}$ in an hour. Presumably the later part of this change corresponds to the conversion of retinaldehyde to retinol; in the early part there may be a contribution from the conversion of metarhodopsin II into pararhodopsin. The peak at 480 nm disappears roughly exponentially with a half-time of about 15 minutes. The changes in the rat (Cone & Brown, 1969; Cone & Cobbs, 1969) are fairly similar to those in the frog, and extremely similar to those in solutions of rat rhodopsin.

Two elegant and simple techniques for studying photolabile pigments in the receptors of intact retinas were developed by E. J. Denton. The first (Denton & Wyllie, 1955) allows single receptors to be studied. The retina is mounted on a microscope slide with the receptors uppermost, and the optical density of single rods is measured, before and after bleaching, by photographing the retina at 36 diameters magnification in monochromatic lights of various wavelengths. Application of this technique to the frog has shown that the mean change of density of a purple rod at 520 nm when it is bleached is 0·66. From this it may be estimated that the density of rhodopsin at 500 nm in such a rod in the unbleached state is about 0·82, i.e. 85% of light of this wavelength falling on the rod is absorbed in rhodopsin.

Denton's second technique (1954, 1955a, b) is to mount the retina folded double, with the receptoral surface outermost in the fold. The margin of the fold then consists of a pile of rods, all lying parallel to each other and perpendicular to the edge. The technique makes use of the dichroism of the rods (see p. 23). The difference of optical density for the two directions of polarization can easily be measured for the pile of rods in the edge of the folded retina, and for the salamander and clawed toad (*Xenopus*) it is found to vary with wavelength almost exactly as does the absorption of light by solutions of the photolabile pigment which can be extracted from the rods: the photolabile pigment accounts for all, or nearly all, the dichroism.

Immediately after bleaching with light, the dichroism which the unbleached rods show in the middle of the visible spectrum disappears, and is replaced by a dichroism in the same direction in the violet and near ultraviolet. During the following half-hour, this dichroism disappears, and then, at least in *Xenopus*, reverses, so that near ultraviolet light polarized with its electric vector along the axis of the rods is preferentially absorbed.

*Spectrophotometry of the retina in the living eye (ophthalmoscopic densitometry).* At the end of the last century, Abelsdorff observed with an ophthalmoscope the bleaching and regeneration of purple pigment in the retinas of living fish (1895) and crocodiles (1898), choosing species in which the retina is backed by a highly reflecting white tapetum lucidum, so that a large fraction of the light striking the retina was returned after passing through it, and a part of this came out through the pupil. Abelsdorff's observations were qualitative only. Brindley & Willmer (1952) extended them to the human eye, a more difficult object of investigation because the retina is backed by heavily pigmented layers which allow only about 1% of light falling on them to return. By measuring the fraction of monochromatic light at various wavelengths reflected, they obtained a lower bound for the density of rhodopsin in the human retina. Subsequently Rushton and his collaborators (see especially Rushton, Campbell, Hagins & Brindley, 1955; Rushton & Henry, 1968) have developed very sensitive photoelectric techniques for measuring the fraction of light incident on the fundus oculi which is reflected, and thus studying the bleaching and regeneration of photolabile pigments in the retina. These techniques have been applied to living and freshly excised eyes, at first of animals with unpigmented or reflecting layers behind the retina, and later also of those with densely pigmented eyes.

Essentially similar techniques, in some respects a little better and in some a little worse, were developed independently by Weale (see especially Weale, 1953b, 1967). Weale's apparatus has, on the whole, been less subtly used than Rushton's, and most of the discoveries in the field were made first by Rushton or his collaborators. But Weale's paper of 1967 provides significant information which could not have been obtained with Rushton's slower densitometer.

In excised eyes of albino rabbits, Hagins (1956) found that sudden bleaching with a flash of less than a millisecond's duration is followed by a decrease in extinction at 486 nm and a concurrent increase in extinction at 400 nm which proceeds exponentially with a half-time of 20 msec at 12°C. The rate-constant follows Arrhenius's law with a heat of activation of 27 kcal/mole, so that at 37°C it should be only about 120 μsec. Hagins suggested that the process concerned was the breakdown of metarhodopsin (i.e., in later terminology, metarhodopsin I). This seems almost certainly correct. It is a good deal faster in the intact rods of the rabbit than in solutions made from cattle retinas, if one may extrapolate from Fig. 6 of Matthews et al. (1963).

In living human eyes, Rushton (1956a) obtained evidence for an orange-coloured intermediate which lasted several minutes after bleaching of the rhodopsin by light. The orange-coloured intermediates found in the isolated retina of the frog by Brindley (1960b) and Bridges (1962b) and in that of the rat by Frank & Dowling (1968) are comparable in stability with that found by Rushton in man, if one takes into account the lower temperatures at which they were examined. These orange intermediates are evidently much too stable to be metarhodopsin I. Very probably they are transient orange (pararhodopsin).

If a brief flash of light is found to bleach a fraction $p$ of the rhodopsin in a retina, then the simple assumption that all the molecules of rhodopsin are alike, and that absorption of light is always followed by bleaching, leads necessarily to the conclusion that a flash $n$ times as bright will bleach a fraction $(1-(1-p)^n)$ of the rhodopsin. If $p$ and $n$ are fairly large, the brighter flash should bleach nearly all the rhodopsin present. For the rabbit's

eye subjected to flashes of light of less than a millisecond's duration, this is not observed (Hagins, 1955). Such flashes bleach a
constant fraction of the rhodopsin present, approximately one-
half, independently of their brightness, provided that this exceeds
a critical value. When half of the rhodopsin has been thus bleached,
another flash delivered a few seconds later will bleach half of the
remainder. Hagins (1958) found that this restoration of about
half the original amount of manifest photolability proceeds
exponentially with a half-time of 20 msec at 8°C. Two possible
explanations of this can be suggested. The first is that the absorption of light by a molecule is not necessarily followed by bleaching.
After a molecule has absorbed a quantum of light it passes into an
excited state from which it may, with about equal probability,
pass along the sequence of thermal reactions that leads to bleaching or return to the ground state. The second explanation of
Hagins's observations makes them analogous to what Hubbard
& St George (1958) found in the rhodopsin system of the squid.
Light not only converts rhodopsin into a substance which we
will for the moment call $x$, but also converts $x$ into rhodopsin
(or a mixture of rhodopsin and isorhodopsin; see p. 22). If
the photosensitivities of $x$ and rhodopsin are equal, then any
bright flash will convert either into a mixture of equal quantities
of the two. If $x$ decomposes thermally, then a brief flash, followed
by a dark interval long enough for it to have disappeared, will
leave half the rhodopsin, of which another flash can then again
bleach half.

These two explanations of the Hagins phenomenon are con-
ceptually distinct, and accepted photochemical theory would
allow either of them to occur without the other, but it is likely
that in fact both are true. The experiments of Hubbard & Kropf
(1958) make it nearly certain that the second suggested process
occurs, the principal reversible photoconversion being that
between rhodopsin and metarhodopsin I. The fact that the
quantum efficiency of bleaching of rhodopsin is less than 1 shows
that a basis for the first exists too, and the spontaneous reversion
of pre-lumi-iodopsin to iodopsin discovered by Hubbard, Bounds
& Yoshizawa (1965) provides an excellent parallel to it.

Analogues of the Hagins phenomenon have been demonstrated
in solutions of frog rhodopsin (Bridges, 1961), in human foveal
after-images (Brindley, 1959a), and in human foveal cone pig-

ments (Rushton & Baker, 1963). For rhodopsin in the living human eye an analogue was found by Rushton (1963*c*), though denied by Ripps & Weale (1969).

*Chemical examination of isolated retinas after illumination.* Further evidence for the similarity of the sequence of thermal reactions that follow absorption of light in solution and in the rod comes from the old experiment of Wald (1935), who found that if frogs' retinas are bleached and extracted with light petroleum or carbon disulphide while they are in the orange or yellow stage, retinaldehyde is obtained, together with a trace of retinol. If extraction is delayed until they have become colourless, the solution contains much more retinol, but no retinaldehyde. From dark-adapted retinas these solvents extract neither retinaldehyde nor retinol.

*The early receptor potential.* Brown & Murakami (1964) found that when a monkey's eye was stimulated with an extremely bright brief flash of light, an electrical response with no detectable latency could be recorded by means of an intraretinal microelectrode. Cone (1964) showed that this 'early receptor potential' could equally well be recorded with large electrodes outside the retina, that its amplitude was directly proportional to the amount of visual pigment bleached, and that in the rat the action spectrum for producing it agreed with that for bleaching rhodopsin. The response is biphasic, the first phase being vitreous-positive and the second phase vitreous-negative. Most investigators have supposed that it is the sum of two components with these polarities; the fast vitreous-positive component is often called $R_1$, and the slower vitreous-negative component $R_2$. When recorded from the isolated retina the early receptor potential retains its normal time-course in sodium-free solutions (Brindley & Gardner-Medwin, 1965; Pak, 1965), and is not much altered by high concentrations of metal-chelating agents, or of $N$-ethyl maleimide, formaldehyde or hydroxylamine (Brindley & Gardner-Medwin, 1966) though it is almost abolished by glutaraldehyde (Arden, Bridges, Ikeda & Siegel, 1966). On heating the retina, the early receptor potential disappears at just that temperature at which the regular orientation of the visual pigment molecules is lost (Cone & Brown, 1967).

Given all these facts, the view that the whole of the early receptor potential is directly due to movements of charge in visual pigment molecules during the photochemical reaction and the earlier members of the chain of thermal reactions that follow it (Lettvin, 1965; Brindley & Gardner-Medwin, 1966; Cone, 1967; Ebrey, 1968) is difficult to escape; Arden & Miller (1968) have argued that such an explanation applies only to $R_1$ and not to $R_2$, but their argument rests on the assumption that formaldehyde and glutaraldehyde cannot influence the kinetics of the thermal decay processes, which seems unreasonable. The hypothesis that $R_2$ depends on the conversion of metarhodopsin I to metarhodopsin II in just the same way that $R_1$ depends on an earlier stage in the thermal decay (most likely pre-lumirhodopsin to lumirhodopsin) gains strong support from the agreement of its time-course and the effect of temperature on it with the kinetics of conversion of metarhodopsin I to metarhodopsin II as determined in the intact retina (Hagins, 1958; Ebrey, 1968). The conversion is one that depends much on pH, and if the hypothesis is right, the rod early receptor potential must show a corresponding dependence on pH. This is not yet known; the effect of pH has been explored (Brindley & Gardner-Medwin, 1966; Pak, Rozzi & Ebrey, 1967) only in the frog, whose early receptor potential is now known to depend mainly on cones (Goldstein, 1967).

Visual pigments are almost certainly contained in the surface membranes of flattened closed or nearly closed sacs (see Fig. 1.8), and the electronmicroscopical appearance of these sacs suggests that opposite surfaces of them face in opposite directions. This raises a problem: where is the asymmetry by which events in the pigment molecules cause electric current to flow at a distance? Brindley & Gardner-Medwin (1966) suggested that the electron-microscopical appearances must be misleading; the molecules must face predominantly in one direction. If so, this would be a specialization almost sure to have some functional significance. But Lettvin (1965) made the suggestion (which Gardner-Medwin and I had not seen when we corrected our proof) that there might be sufficient asymmetry arising from the connexion of the interior of some (not all) of the receptor sacs to extracellular space. Lettvin's suggestion gains support from the discovery (Goldstein, 1967; Goldstein & Berson, 1969) that the early receptor potential in the frog and in man depends mainly on cones, though rods are more numerous than cones in both species. As direct electron microscopy suggests, and Cohen (1968) elegantly supported by experiments on the penetration of lanthanum, connexions to extracellular space are usual in the sacs of cones and rare in those of rods.

The early receptor potentials of rods and of cones are very similar in time-course in the frog (Goldstein, 1968, and personal communication). On the interpretation adopted above, this implies that the kinetics of thermal decay of pre-lumirhodopsin, lumirhodopsin and metarhodopsin I and of their cone analogues are closely similar. The available information about the iodopsin analogues (see p. 40) is compatible with such similarity, though not sufficient to establish it.

*Rapid measurement of the pH of rod suspensions and rhodopsin solutions.* Falk & Fatt (1966) have measured the pH of frog rhodopsin solutions and suspensions of outer segments of frog rods with a platinum electrode, and found that during the first few milliseconds after flash illumination of the solution or suspension there is a rapid uptake of hydrogen ions whose time-course and relation to the pH of the medium are what would be expected from the conversion of metarhodopsin I to metarhodopsin II. This supports other indications that the processes in rods differ little from those in solutions of rhodopsin.

### The regeneration of rhodopsin within the rods

Boll observed in 1876 that retinas removed from animals that had been exposed to bright light were colourless, but that, if the animals were kept in the dark for some time before they were killed, the purple colour was restored. Ewald & Kühne (1878c) estimated the time required for such regeneration in the living eye of the rabbit at 35 minutes. In the frog, Kühne (1878a) showed that regeneration in the isolated retina is slow and very incomplete, but becomes faster and much more complete if the bleached retina is placed in contact with the pigment epithelium of the eye from which it was removed. Almost certainly this difference is due to the retinaldehyde isomerase of Hubbard (1956), which in the frog is present in the pigment epithelium and absent in the retina. In the isolated mammalian retina there is, under most conditions, almost no regeneration (see, for example, Weinstein, Hobson & Dowling, 1967); but Cone & Brown (1969) found that it occurs quite efficiently if the retina is incubated at 37°C in a very small volume of solution.

The time-course of regeneration of rhodopsin in the living eye was first investigated by Tansley (1931), who killed rats at various stages of dark-adaptation, extracted the rhodopsin from their retinas with digitonin, and measured the amount of it in the solutions colorimetrically. The continuous observation of rhodopsin in the retinas of living animals by measurement of the light reflected from the fundus oculi has made it possible to study regeneration in a number of species, including man, with greater accuracy and much less labour than when the rhodopsin had to be extracted before it could be measured. Figure 1.9 shows the course of regeneration after complete bleaching in man.

Rushton (1957) has shown that in man a blue and a yellow light, adjusted to be of equal scotopic brightness and therefore equally effective in bleaching rhodopsin, but very differently absorbed by retinaldehyde (about 10 : 1), produce the same reduction in

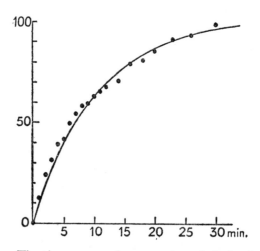

FIG. 1.9. The time-course of regeneration of rhodopsin in the living human retina. *Ordinates:* percentage regenerated. The curve drawn through the experimental points is exponential with a half-return period of 7 minutes (from Rushton, Campbell, Hagins & Brindley, 1955).

the amount of rhodopsin in the retina at equilibrium in the light, and the same rate of regeneration in the dark after bleaching, as measured by ophthalmoscopic densitometry. Similar results have been obtained by Lewis (1957) in the albino rat. Thus it appears that photoisomerization of free retinaldehyde is not important for the regeneration of rhodopsin in the mammalian eye. The photoisomerization of all-*trans* retinaldehyde still attached to opsin (especially pararhodopsin, i.e. transient orange) may however make some contribution.

## Movement of carotenoids between retina and pigment layers

Jancso & Jancso (1936) showed that after rhodopsin is bleached in the living eye, retinol appears in the pigment epithelium and choroid. The total amount of vitamin A in the eye as free retinol, esterified retinol, free retinaldehyde and combined retinaldehyde (chiefly rhodopsin) remains practically constant throughout the bleaching and regeneration of rhodopsin (Hubbard & Colman, 1959; Dowling, 1960), but much of the retinol formed as a result of bleaching is conveyed to the pigment layers, and much of the retinol required during regeneration is taken from the pigment layers.

The greater part of the ocular retinol, both in the retina and in the pigment epithelium and choroid, is not free but esterified with long-chain fatty acids (Krinsky, 1958*b*). Part of this esterified retinol is in the 11-*cis* configuration; the eye is the only part of the body that is known to contain 11-*cis* retinol. Hubbard & Dowling (1962) have found that bleaching of rhodopsin increases the amount of all-*trans* retinol ester in the frog's eye without affecting the amount of 11-*cis*. During regeneration of rhodopsin, all-*trans* and 11-*cis* esters decrease, the ratio of their concentrations remaining approximately constant. After regeneration is complete, the total amount of retinol ester remains constant, but the proportion of 11-*cis* slowly rises during many hours. It is not known whether the conversion of all-*trans* retinol into 11-*cis* retinol ester is direct or goes through free retinol or free retinaldehyde; the last seems the most likely, in view of the known properties of retinaldehyde isomerase.

## Photolabile pigments of the retina other than rhodopsin

*The range of pigments contained in rods*

On the basis of investigations by Köttgen & Abelsdorff (1896) and himself on a number of species, Wald (1939) suggested the generalization that the rods of land vertebrates and marine fish contain rhodopsin, those of freshwater vertebrates a distinct pigment called porphyropsin, and those of amphibia and euryhaline fish both rhodopsin and porphyropsin. Rhodopsin was described as absorbing maximally at 500 nm and containing retinaldehyde as the carotenoid part of its molecule, and porphyropsin as absorbing maximally at 522 nm and containing 3-dehydro-retinaldehyde.

As a rough guide to the distribution of rod pigments based on retinaldehyde and 3-dehydroretinaldehyde this rule is still useful, though there are exceptions to it: for example several purely freshwater fish have two pigments, one derived from retinaldehyde and the other from 3-dehydroretinaldehyde (see Bridges, 1965). As a guide to the absorption spectra, Wald's generalization can no longer be accepted. Photolabile pigments with absorption maxima varying over the range 462 to 550 nm have been extracted from rod-rich vertebrate retinas, and probably all of these are rod pigments. The distribution within this range is not haphazard. All mammals so far investigated have pigments with absorption maxima within four or five millimicrons of 498 nm. Most freshwater fish agree with Wald's generalization to the extent of having at least one pigment with its absorption maximum between 510 and 550 nm, though they may possess other pigments whose maxima fall outside this range. Many marine fish that live in deep water have pigments whose maxima lie between 475 and 490 nm, as Denton & Warren (1957) first noted on the basis of measurements on whole retinas, and Munz (1958) confirmed by extraction.

Bridges (1965) and Dartnall & Lythgoe (1965) independently noticed that the wavelengths of maximum absorption of retinaldehyde pigments from fish tend to be clustered around a small number of regions in the spectrum. The clustering, though not very strict (Fig. 1.10), is certainly more than can be attributed to chance. The retinaldehyde pigments from amphibia, reptiles, birds and mammals show a slighter tendency to fall into the same clusters. 3-dehydroretinaldehyde pigments fall into a different

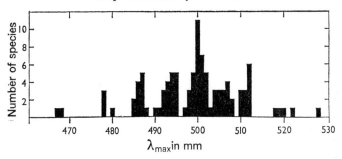

Fig. 1.10. The distribution of absorption maxima of retinaldehyde pigments from teleost fishes (from Dartnall & Lythgoe, 1965).

series of clusters. It seems probable that the reason for this clustering is chemical rather than functional.

Dartnall (1952) pointed out that the positive parts of the difference spectra of a number of photolabile pigments which had been extracted from the retinas of vertebrates could be superimposed on each other when plotted against frequency and shifted by suitable amounts along the axis of frequency. He argued that this property of difference spectra would be valid also for the absorption spectra of the pigments, and suggested that it might hold for other visual pigments still undiscovered. The original comparison is shown in Fig. 1.11.

Dartnall's relation seems to be widely applicable; of the very many photolabile pigments that have been extracted from the retina with digitonin since it was first put forward, none is known to disobey it substantially, but there may be a small departure from superimposability when retinaldehyde and 3-dehydro-retinaldehyde pigments are compared (Bridges, 1965). The colour-matching properties of the human eye make it very probable (on my calculations of the best-fitting linear transformations of the spectral mixture curves) that of the three receptive pigments of human cones two (the blue-sensitive and green-sensitive, called 'cyanolabe' and 'chlorolabe' by Rushton,) are closely super-imposable on Dartnall's standard curve; but the third (the red-sensitive, called 'erythrolabe') deviates substantially from it.

*Cone pigments studied in solution*

To extract cone pigments with the agents that easily extract rhodopsin and other rod pigments seems to be curiously difficult.

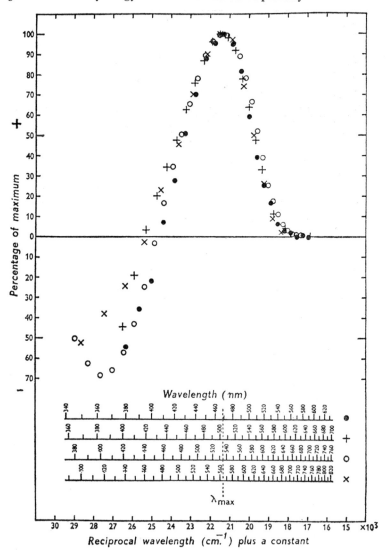

FIG. 1.11. Comparison of the difference spectra of four retinal pigments, plotted against frequency. The spectra have been shifted along the horizontal axis so that their maxima coincide. X, iodopsin; O, tench porphyropsin; +, frog rhodopsin; ●, the red-insensitive pigment of the tench, which absorbs maximally at 467 nm. Appropriate wavelength scales are shown below (from Dartnall, 1952).

No convincing report of the extraction of a photolabile pigment from mammalian, reptilian or amphibian cones has been published, though I know of careful and thorough attempts to extract them that have been unsuccessful, some published (e.g. Bliss, 1949) and others unpublished. Yet cones do contain photolabile pigments, as was made likely when Wald (1937) discovered iodopsin (see below), almost certain when Rushton (1955) demonstrated light-induced changes in the reflectance of the human fovea, and absolutely certain when Marks (1963, 1965), Liebman & Entine (1964), Marks, Dobelle & MacNichol (1964) and Brown & Wald (1964), succeeded in making direct spectrophotometric measurements on single cones.

*Iodopsin.* Birds are the only vertebrates from which cone pigments have (very probably) been extracted. The retinas of diurnal birds contain rods and cones, the cones greatly predominating (Schultze, 1866). In the domestic fowl, in which, according to Schultze's drawings, the cones outnumber the rods by about six to one, Wald (1937) found that digitonin-extracts of whole retinas contained two photolabile substances. The more abundant of them had the properties of rhodopsin. The less abundant absorbed maximally at considerably longer wavelengths; Wald named it iodopsin, and suggested that it was derived from the cones. The two pigments have been further studied by Bliss (1946) and Wald, Brown & Smith (1955). The latter authors were able to synthesize iodopsin by adding 11-*cis* retinaldehyde to bleached extracts of rod and cone outer segments separated from other retinal tissue by centrifugation. If excess 11-*cis* retinaldehyde is added, rhodopsin is also formed, but some 500 times more slowly. This differential synthesis and the possibility of bleaching the iodopsin alone in a solution containing both pigments by using red light provide two independent means of estimating the absorption spectrum of iodopsin, the estimates obtained by the two methods agree very well.

The proportion of iodopsin to rhodopsin in the retinal extracts of Wald, Brown and Smith, measured as the ratio of maximum extinctions, is about 1 : 1·5. Thus the amount of iodopsin extracted from each cone is only about one-tenth of the amount of rhodopsin extracted from each rod, assuming that all the fowl's cones contain iodopsin and all its rods rhodopsin. We cannot be certain whether this represents a difference in the amounts present in the receptors or a difference in ease of extraction; but the approximate constancy of the rhodopsin–iodopsin ratio when three different techniques of extraction are used favours the former alternative.

The sequence of thermal reactions that follows the illumination of iodopsin has been studied in solution by Hubbard & Kropf (1959) and Yoshizawa & Wald (1967). It is mainly similar to the sequence for rhodopsin, though no counterpart in the iodopsin sequence to metarhodopsin II or pararhodopsin has yet been reported.

*Other cone pigments studied in solution.* The pigeon (Bridges, 1962a) has two pigments that seem to be analogous to the two that Wald found in the domestic fowl, and it is likely that the more abundant comes from rods and the less abundant from cones.

From the retina of the grey squirrel Dartnall (1960) extracted a photolabile pigment which he believed to be a cone pigment because the grey squirrel was said by some histologists to have no rods; but the histological observation was probably an error (Cohen, 1964), and Dartnall's pigment is probably that of the grey squirrel's rods.

## Cone pigments in the living retina

On the evidence of its power to discriminate colours, the rod-free foveal part of the human retina must contain at least three receptive pigments, and probably contains only three. If there are only three pigments, we can infer from the colour-matching properties of the eye (see p. 222) that the action spectrum for producing, in any one pigment, the photochemical change that provokes transmission of a message to the brain must be a linear function of three quantities (the spectral mixture functions) that can be accurately measured. There is, however, no completely certain method of determining the coefficients in the linear function. Thus the colour-matching properties of the eye, though they set clear bounds to what the action spectra of foveal receptive pigments may be, do not suffice to determine them exactly.

Rushton (1955, 1963a, b, c; 1965a, b) has detected, by ophthalmoscopic densitometry, two photolabile substances in the foveal part of the human retina, called 'erythrolabe' and 'chlorolabe', whose properties agree closely with what may be inferred about two of the three pigments required to explain the colour-matching properties of the eye. The third postulated pigment, called 'cyanolabe', should on the colour-matching evidence absorb very little light at wavelengths longer than about 500 nm. Rushton's measurements are for technical reasons restricted to such wavelengths; they would not be expected to detect cyanolabe, and do not.

Investigation of the properties of chlorolabe and erythrolabe by ophthalmoscopic densitometry is facilitated by using colour-blind subjects. In protanopic ('red-blind') people, who might be expected, from the confusions of colour that they make, to lack one of the normal receptive pigments, Rushton's technique detects only one photolabile substance, chlorolabe, in the foveal retina, i.e. the difference spectrum on bleaching is independent of the colour of the light used to bleach (see Fig. 1.12a). The difference spectrum of chlorolabe and the action spectrum for bleaching it agree well with the spectral sensitivity curve of the protanope. By the same criteria the deuteranope ('green-blind') also has only one photolabile pigment absorbing appreciably at wavelengths greater than 500 nm (see Fig. 1.12b). The deuteranope's pigment, erythrolabe, differs from the protanope's pigment, and its difference spectrum agrees with the deuteranope's spectral sensitivity curve.

In the normal fovea the shape of the difference spectrum varies with the wavelength of the light used for bleaching, so that there must be at least two photolabile pigments present that absorb in the relevant range of the spectrum. On the evidence of the colour matches made by normal and colour-blind subjects it has long been probable (see p. 228) that the visual pigments of the normal cones are the same as those of protanopic and deuteranopic cones, i.e. that there are in the normal fovea cones containing the pigment that Rushton found in the protanope and called 'chlorolabe', and also cones containing the pigment that he found in the deuteranope and called 'erythrolabe'. Rushton's observations on the normal fovea are consistent with this supposition, though they do not firmly establish it. The time-courses of regeneration of chlorolabe and erythrolabe in colour-blind eyes, and of the presumably identical pigments in normal eyes, are, within experimental error, the same. They are exponential, with a time constant

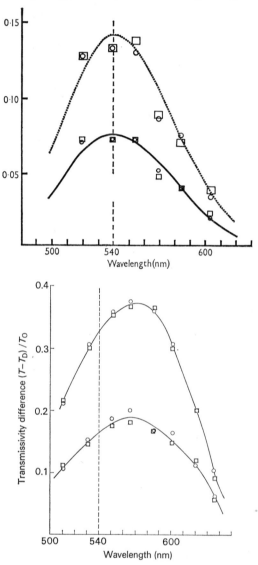

FIG. 1.12. Difference spectra by ophthalmoscopic densitometry for the protanopic (*a*) and deuteranopic (*b*) fovea. The upper curves are for full bleaches with white light, the lower for partial bleaches with red light (circles) or blue-green light (squares). The agreement of the squares with the circles indicates that in each case only one photolabile pigment is present (from Rushton, 1963*a* & 1965*a*).

of about 2 minutes (see Fig. 1.13). They agree closely with foveal dark-adaptation curves on the assumption that bleaching each 9% of the pigment halves the sensitivity of the eye.

## Cone pigments in the isolated retina

Cone pigments can easily be studied by direct spectrophotometric measurements on a region of isolated retina containing some

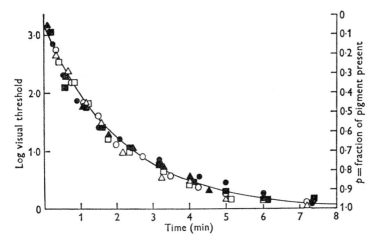

FIG. 1.13. The time-course of regeneration of erythrolabe in the dark in the deuteranopic fovea, compared with the time-course of foveal dark adaptation in the same subject. The arbitrary vertical scaling assumes that sensitivity is halved by the bleaching of each 9% of the pigment present in the fully dark-adapted state. It will be noticed that this assumption becomes absurd in the extreme case where all the pigment is bleached; experimentally, it has been tested only up to about 90% bleaching (from Rushton, 1965*b*).

thousands of cones, for example the foveal region of a primate retina (Brindley, 1963*a*; Brown & Wald, 1963). This technique is more accurate than ophthalmoscopic densitometry, and less subject to uncertainties due to stray light. Brown and Wald obtained evidence by it that human and monkey chlorolabe and erythrolabe are, like rhodopsin, derivatives of the 11-*cis* isomer of retinaldehyde: they regenerate, in the isolated retina, only if this is added. The amount regenerated is surprisingly small; Brown

and Wald never succeeded in restoring, by adding 11-*cis* retinaldehyde, more than half the density originally present, and they
have given no evidence that the regenerated pigments resemble
the original ones in their properties. However, the inference that
chlorolabe and erythrolabe are 11-*cis* retinaldehyde derivatives is
probably correct, for it gains independent support from the
observation of Rushton (1968) that in a parafoveal region of the
retina where rods and cones are similar in abundance the dark-
adaptation of rods is impeded by bleaching the cone pigments.
This is most simply explained by supposing that they compete for
a common ingredient needed for regeneration, very probably
11-*cis* retinaldehyde.

Spectrophotometry of the isolated retina is an especially favourable technique for detecting small amounts of red-sensitive pigments in the presence of very large amounts of rhodopsin. In
conjunction with serial bleaching by red light it easily yields
estimates (Brindley, 1963*a*) of the difference spectra of red-
sensitive pigments in the retinas of frogs and cattle; these are very
probably cone pigments, though cones are not very abundant in
these retinas, and I have been unable to detect cone pigments
in digitonin extracts from them.

*Spectrophotometry of single cones*

It is more difficult to investigate single cones than groups of
them, chiefly because it is only by selecting a photocell with exceptionally high quantum efficiency and using it optimally that the
measurements can be made without bleaching a good deal of the
pigment. The technical difficulties have been overcome by Marks
(1963, 1965), and Liebman and Entine (1964), who examined the
relatively large cones of fishes freely suspended in an aqueous
solution or gel, and by Marks, Dobelle & MacNichol (1964) and
Brown & Wald (1964), who examined the smaller cones of man
and monkeys in whole-mounts of the retina illuminated from its
vitreal surface. The observed optical densities of bleachable pigment are very low, from 0·01 to 0·03, but there are many factors
that would tend to make them lower than the effective densities
in the living eye, especially stray light reaching the recording
photocell without having passed through the cone, and distortion
of the retina whereby the light does not pass along the whole
length of the cone (see also p. 160). The observed densities are

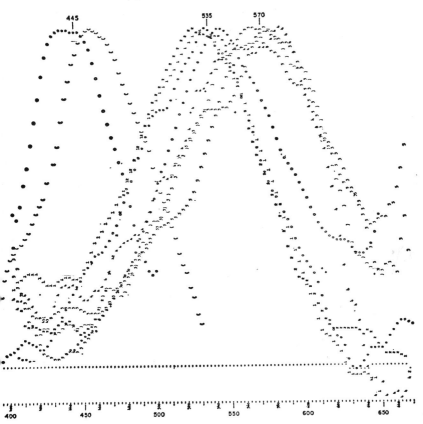

FIG. 1.14. Difference spectra of single human and simian cones. Brackets represent cones from man, other symbols cones from *Macaca nemestrina* and *Macaca mulatta* (from Marks, Dobelle & MacNichol, 1964).

lower by at least a factor of 2 than those that Rushton directly observed by ophthalmoscopic densitometry, though this technique also must substantially underestimate the true density. Thus they do not disprove inferences from sensory experiments (see p. 221) that the optical density of receptive pigment along a foveal cone is between 0·5 and 1, though they give these inferences no support.

Difference spectra for single human and simian cones are shown in Fig. 1.14. Most of them agree roughly with either chlorolabe or erythrolabe, as far as the rather poor accuracy of the measurements permits the comparison. Some have maxima near 450 nm, as was to be expected from the colour-matching properties of the eye; these 'cyanolabe' cones seem to be scarcer than the chlorolabe and erythrolabe cones.

# 2

# THE SIGNALS IN RODS AND CONES

## Electrical activity of rods and cones

AT least until 1962 it could reasonably be doubted whether rods and cones produced any electrical responses; but experiments published in and since that year prove clearly that they do. Part of the evidence comes from the identification, by means of extra-cellular microelectrodes, of the contribution that they make to the electroretinogram; this will be considered mainly in Chapter 3. Part comes from electrical recording from within cones themselves, and this will be considered at once.

The first claim to record electrically from within cones was erroneous, but important. Svaetichin (1953, 1956) inserted fine microelectrodes into the retinas of fish, and was able to record responses of large amplitude (tens of millivolts) and simple time-course. The potential altered quickly during the 100–200 milliseconds after the beginning of illumination, remained nearly steady during steady illumination, and returned quickly to its resting value during the 100–200 milliseconds after illumination ceased. Svaetichin argued that these responses came from cones, but it was shown by Tomita (1957) and MacNichol & Svaetichin (1958) that their sources are in the inner nuclear layer. They are now usually called s-potentials, and will be considered in Chapter 3.

It has since become clear that responses very similar in time-course to s-potentials but smaller in amplitude can be recorded from within cones. The first evidence for such cone responses came from the observation of Oikawa, Ogawa & Motokawa (1959) that s-potentials of two kinds could be recorded from the isolated retina of the carp. The first gave responses whose amplitude was roughly proportional to the area of a small stimulating spot of light centred on the electrode. They could be found with electrodes whose tips were as large as 3 μm, and the sites from which they were recorded could be proved by marking with silver to be in the inner nuclear or outer plexiform layer. The second kind gave responses whose amplitude was nearly independent of the size of the stimulating spot of light. They could be recorded only with electrodes whose tips were less than 0·2 μm in diameter, and for this reason attempts to mark the sites with silver were unsuccessful. Oikawa *et al.* suggested that responses of their second kind came from cones, and the subsequent work of Tomita (1965), Tomita, Kaneko, Murakami & Pautler (1967), Kaneko & Hashimoto (1967), and Werblin & Dowling (1969) has shown that the suggestion was correct. The technical difficulties of marking the recording sites have been overcome; in the carp and in *Necturus* the tip of the electrode is in the inner segment of a cone when the responses are recorded. The inner segment has a resting potential of from $-30$ to $-40$ mV as measured, and the response to light is a hyperpolarizing one.

Toyoda, Nosaki & Tomita (1969) found in *Necturus* cone inner segments and gecko rod outer segments that during the response the resistance of the cell membrane does not fall, as during excitatory or inhibitory postsynaptic potentials in nerve cells; it rises substantially. The increase in membrane resistance is not secondary to the hyperpolarization, for passively hyperpolarizing the cell by passing current into it does not affect the membrane resistance. The ionic basis of the electrical response of cones and perhaps rods could, until the observations of Toyoda *et al.* that the membrane resistance increases, have been supposed to be similar to that of an inhibitory post-synaptic potential, i.e. principally an increase in the permeability to potassium and chloride ions. An appropriate effect of light on isolated outer segments of rods had been reported: Etingov, Shukolyukov & Leontyev (1964) and Bonting and Bangham (1967) found that rods suspended in sucrose solutions lost potassium faster if they were illuminated than if they were kept in darkness, and Daemen & Bonting (1968) showed that the addition of retinaldehyde (11-*cis* or all-*trans*) to the bathing solution imitates this action of light. But such an effect cannot explain the increase in membrane resistance. Toyoda *et al.* suggest that the primary action of light on receptors is to make their membranes less permeable to ions (presumably chiefly sodium) whose passive entry tends to depolarize the membranes. The suggestion explains all the microelectrode observations, and has its counterpart in the finding of Bonting & Bangham (1967) that the augmentation of potassium loss from rods during illumination is accompanied by a smaller but clear diminution of sodium loss. Nevertheless, I am reluctant to accept it as part of the mechanism of vision, for it seems difficult to explain by it the sensitivity of a rod to a single quantum and of a cone to no more than half a dozen quanta (see pp. 186–189).

Replacement of sodium by an inactive cation of low membrane permeance, such as choline, rapidly and reversibly abolishes or greatly reduces the whole of the electroretinogram of the isolated retina (Furukawa & Hanawa, 1955; Hamasaki, 1963). The component (called 'distal PIII'; see p. 69) that is believed to be generated by rods and cones disappears with the rest. Ouabain, whose well known action is to block the active pumping of sodium and potassium across cell membranes, likewise abolishes or greatly reduces the whole electroretinogram, but only after it has acted for many minutes (Ostrovskiĭ & Dettmar, 1968). These two indications that distal PIII, like the rest of the electroretinogram, depends somehow on sodium, have been much refined by Arden & Ernst (1969). They found that ouabain slowly abolishes PIII (isolated, in the pigeon's retina, by means of a calcium-free high-phosphate medium), but when the cone constituent of PIII has been abolished by ouabain it can be restored with normal polarity by increasing the external sodium concentration, or with inverted polarity by decreasing the external sodium concentration. These procedures do not restore the rod constituent of PIII. Arden and Ernst argue (independently of the concurring argument of Toyoda *et al.*) that in cones the action of light is to diminish the permeability of the surface membrane to cations. They suggest (but give no clear arguments) that in rods light acts on electrogenic pumping of ions. The permeability that in cones is diminished by light would, on the hypothesis of Arden and Ernst, be to potassium as well as sodium, since they found that in ouabain-poisoned pigeon retina potassium acts like sodium in restoring a cone PIII.

Although Toyoda, Nosaki & Tomita (1969) found an increase in membrane resistance on illumination in gecko rods and *Necturus* cones, and Arden & Ernst (1969) provided indirect evidence for such an increase in pigeon cones, the accurate measurements of Falk & Fatt (1968*a*, *b*) on packed suspensions of frog rod outer segments revealed no such increase; on the contrary, there is an apparent decrease, though Falk and Fatt are careful to point out that this might not be real, but might depend on an effect of light on the volume of the rods.

## Grounds for supposing that the electrical activity of rods and cones constitutes a link in the chain of conduction of the signal

The responses that have been recorded from within cones are similar in time-course to the distal PIII component of the electroretinogram, and the polarity of this component (very probably receptoral) is what would be expected if the cone electrical responses were actual signals, i.e. were generated by movement of ions across the membrane of some outer part of the cone, and propagated by longitudinal passive flow of current to parts nearer to the synaptic pedicle. It is natural to suppose that the cone electrical responses (and presumably also similar responses of rods) are signals. This does not at once imply that they signal all the way from the outer segment to the cone pedicle. One could suppose that the movements of ions across the cell membrane that generate the electrical responses occur in the inner segment, and that absorption of light in the outer segment is signalled to the inner segment by diffusion of a chemical substance, by a mechanical pull or push, or by propagation of a dislocation in a crystal lattice. But these suppositions were never very plausible, and they are disproved by the recent experiments of Penn & Hagins (1969), who in small slices of isolated rat retina analysed the PIII part of the electroretinogram with extracellular microelectrodes in much finer detail than any of their forerunners. The experiments of Penn and Hagins are a combination and refinement (with very great improvement in technique) of those of Murakami & Kaneko (1966) and Brindley (1956a), discussed on pp. 66 and 69. The response to light was recorded between two microelectrodes, one with its tip a little beyond the outer extremities of the rods and the other inserted between the rods to an accurately controlled depth (see Fig. 2.1). Penn and Hagins also recorded the steady difference of potential in the dark between such electrodes, and measured the conductivity of the retina at various points for current parallel to the axes of the rods. They could therefore infer

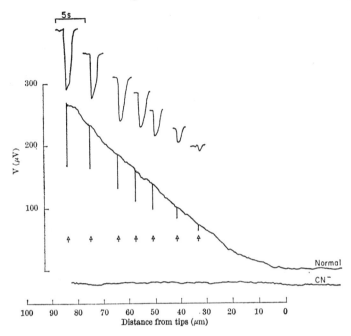

FIG. 2.1. Electrical activity of rods in a small slice of rat retina. The curves labelled 'normal' and 'CN⁻' show the standing difference of potential in the dark (without and with poisoning by cyanide respectively) between a microelectrode inserted from the outer surface to the depth shown on the horizontal axis and a reference electrode at the tips of the outer segments. Each of the separate curves above shows the response to a very brief (1 μsec) flash of light recorded in the same way and displayed with the same scale of potential but with time on the horizontal axis (from Penn & Hagins, 1969).

the extracellular current in this direction, both in the dark and during the response, and from the distribution of extracellular current calculate the membrane current across the rod membranes, since it is the first derivative of the extracellular current with respect to distance (see Fig. 2.2). The amount of charge transferred across the membrane during a response is $1 \cdot 5 \times 10^7$ electronic charges per outer segment, for a stimulus from which it is estimated that 250 quanta are absorbed per flash. The distribution of response membrane current along the rods is similar to

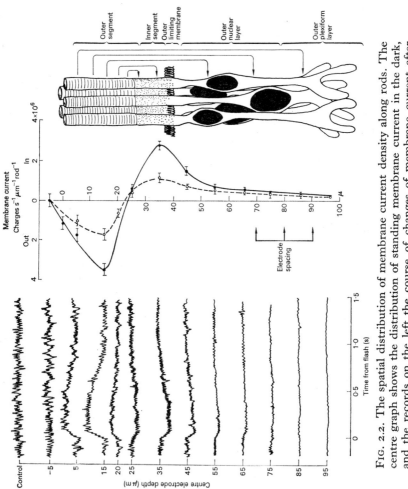

FIG. 2.2. The spatial distribution of membrane current density along rods. The centre graph shows the distribution of standing membrane current in the dark, and the records on the left the course of changes of membrane current after a brief flash, recorded at the depths shown on the vertical axis (from Penn &

that of the dark membrane current except that it is opposite in sign. For the stimuli used by Penn and Hagins, the response membrane current was always smaller than the dark membrane current. Extrapolation of their results suggests that it may never exceed the dark membrane current for any stimulus, as some plausible notions about its origin predict.

It is very probable that propagation of the hyperpolarization along a rod or cone is purely passive; it is not assisted by any regenerative process analogous to the sodium carrier mechanism of nerve fibres. For cones, this can be inferred from the fact that the electrical response is continuously graded according to the strength of the stimulus, as the records of Tomita *et al.* (1967) show. For rods, the records of membrane current reproduced in Fig. 2.2 make it nearly certain. The passivity of conduction has one interesting implication that accords well with old observations. The electrical response of a cone is of small amplitude; it is not more than 5 mV in the inner segment for the rather bright lights ($2 \times 10^{13}$ quanta cm$^{-2}$ sec$^{-1}$) used by Tomita and his colleagues. For dimmer but still easily visible lights it will be smaller, perhaps very much so, and it will be even smaller in the synaptic pedicle than in the inner segment. The experiments of Penn and Hagins suggest that the situation in rods is similar. We should therefore expect that very weak applied electric currents running in the direction of the inner parts of rod and cone cells will easily imitate this small signal and cause sensations of light. They do so; the preferred directions of stimulating current are those of the inner parts of rods and cones (see Fig. 6.6), and the electrical threshold for short pulses is only $8 \cdot 3 \times 10^{-9}$ coulomb cm$^{-2}$, i.e. less than $10^4$ electronic charges per rod or foveal cone (Brindley, 1955a). The change in membrane potential produced by such a pulse, assuming a membrane capacity of $1 \cdot 5 \times 10^{-5}$ μF per rod (Hagins, 1965) will be less than 100 μV; how much less depends on the relation between the intracellular and extracellular radial resistance of the layer of rods and cones, which is quite unknown. The effect of the polarity of a long pulse of current (Helmholtz, 1896, p. 247) is appropriate: current flowing from vitreous to sclera, which should hyperpolarize the synaptic terminals of rods and cones, causes seen light to look brighter; current flowing from sclera to vitreous, which should depolarize the terminals, causes it to look dimmer.

c

## Objections to the hypothesis of electrical signals

The hypothesis that the electrical response of rods and cones is the signal that they transmit is very attractive. There are, however, objections to it, which were very clearly set out by Cone & Ebrey (1965). When, in a rat, PIII, or perhaps distal PIII, is isolated by obstructing the retinal artery, it is found to have the following properties:

(1) it clearly signals the beginning of a period of illumination, but gives little or no immediate indication of its cessation;

(2) it is very insensitive to incremental stimuli on bright backgrounds;

(3) it is very insensitive to flicker at 3 c/sec.

But the retinal ganglion cells, in the rat as in other mammals, are as sensitive to the cessation of a period of illumination as to its beginning, and respond very well to incremental stimuli and to flicker at 3 c/sec. Thus if these properties are absolute and are true reflections of properties of the electrical responses of normal receptors (in this case probably rods, since the rat has very few cones), we are forced to conclude that the electrical responses are not the signals (or not the only signals) that transmit information from the outer segments to the nerve cells of the retina. Cone and Ebrey argued convincingly that the three troublesome properties of PIII are not artifacts from insufficiency of the blood supply to receptors when the retinal artery is obstructed; so if we are to rescue the hypothesis that the electrical responses of receptors are the signals, we must rely on the supposition that the three properties are not absolute, i.e. that the PIII responses to cessation of illumination, to incremental stimuli, and to flicker, are not absent, but merely so small as to be lost in noise under ordinary conditions of recording. If so, it might be possible to detect them by averaging many responses.

Another difficulty in accepting PIII or distal PIII as the receptor signal comes from the observation, made by many experimenters but most clearly by Cone (1963), that there is a wide range of weak stimuli, roughly between one quantum absorbed per hundred rods and three or four quanta absorbed per rod in the rat, that produce a b-wave (i.e. signal of opposite polarity to PIII), but no detectable preceding a-wave (i.e. signal of the polarity of PIII). Here again, we can rescue the hypothesis that PIII is the receptor signal if the a-wave for weak stimuli is not really absent, but merely too small to have been yet detected. There is nothing unacceptable in this, though it implies that rods work very near to the limits set by thermal noise.

These objections to the hypothesis of electrical signals in receptors deserve to be kept in mind; they will not entirely lose their force until receptoral responses to cessation of illumination, to small increments, to flicker, and to very weak flashes have been detected. Nevertheless, I think it probable that in the end (but perhaps only after techniques have been greatly improved) such responses will be detected, and the hypothesis of electrical signals established as fact.

## Indications that the receptor signal is proportional to the stimulus

It has long been widely supposed that in general the messages from sense organs are roughly linearly related to the logarithm of the strength of the stimulus. There are, however, several indirect indications from sensory experiments that the signals transmitted by rods and cones may not follow this rule, but may, over a substantial range, be directly proportional to the intensity of light falling on them.

It might at first sight be supposed that one such indication is Riccó's law, i.e. the law that within certain limits the effectiveness of a given amount of light in reaching visual threshold is independent of its spatial distribution on the retina (see p. 163). This seems to show that the effect of multiplying the illumination of each rod by any number $n$ can be compensated for by dividing the number of rods illuminated by $n$. Such a compensation could be much more economically explained if responses were proportional to stimuli than if they were not. But in fact the classical demonstrations of Riccó's law at the absolute threshold of the human dark-adapted eye tell us nothing about the effect of varying the illumination of a rod, since at absolute threshold most rods receive no light, a minority receive one quantum, and practically none receive more than one quantum. Riccó's law under such circumstances merely shows that the visual pathway, in deciding whether a flash has been received, counts how many rods have received a quantum without much regard for their position.

Riccó's law also holds, however, under conditions that are far from absolute threshold—conditions where each rod or cone receives many quanta per stimulus. Such conditions are shown in Fig. 8.18 (Brindley, 1954a) for blue-sensitive cones, and Fig. 7.2 (Barlow, 1958) for rods. Another clear case is to be found in the experiments of Alpern, Rushton and Torii (1970), where the effectiveness of a flash in causing metacontrast is found to depend on the number of quanta absorbed in rods, independently of their distribution between rods, up to as many as 100 quanta per rod per flash. Cases like these suggest, but do not prove, that over most of the working range of rods and an important part of the working range of cones, the signals are proportional to the stimulus. Direct investigation of the question might be attempted by intracellular recording, by examining the earliest part of the normal electro-retinogram, or by isolating PIII by chemical agents. However, all three methods are less than fully satisfactory. By intracellular recording, receptor signals large enough to be measured can be obtained only with strong stimuli, in the upper part of the range in which proportionality is to be expected or even outside that range. Analysis of the normal electroretinogram may reveal properties of the earliest part of the receptor signal, but cannot disentangle its later parts. Chemical isolation of distal PIII is open to the criticism that the receptor signal that is thus revealed may not be normal and may not be uncontaminated by other retinal activity, especially 'proximal PIII' (see p. 69).

# 3

# THE NERVE CELLS OF THE RETINA

## The connexions between the nerve cells

THE broad classes of nerve cells in the retina are the same in all or nearly all vertebrates. Light-microscopical techniques, especially the Golgi method, sufficed during the last decades of the nineteenth century to reveal the shapes of the cells and some of their interconnexions. Cajal (1894) gave a very thorough and clear account of such observations, mainly his own. Techniques very similar to those of Cajal and his contemporaries are still useful in investigating retinal structure, and Brown & Major (1966), Leicester & Stone (1967) and Boycott & Dowling (1969) have made new discoveries with them. However, the most striking recent advances in our knowledge of the connexions between retinal nerve cells have come from electron microscopy. This is not an appropriate place to review so large a subject fully, but Fig. 3.1 summarizes the conclusions that Dowling and Boycott (1966) draw from electron-microscopical observations on human and simian retinas.

Each rod (R) ends in a terminal spherule which is invaginated by the dendrites of several cells (in man usually 4 or 5, according to Missotten, Appelmans & Michiels, 1963). Some of these cells are rod bipolars (RB), others horizontal cells (H). Many synaptic vesicles, and a special structure called a synaptic ribbon (see Fig. 3.2), are found close to the junction on the presynaptic (rod) side.

Each cone (C) has a terminal pedicle, which is invaginated by many sets of three dendrites. Missotten et al. counted 25 of these 'triads' invaginating a typical human cone pedicle. Dowling and Boycott think it likely that the middle dendrite of each set of three comes from a midget bipolar (MB) and the outer two from different horizontal cells. Each synapse between a cone pedicle and a triad has, close to the junction on the presynaptic (cone) side, a synaptic

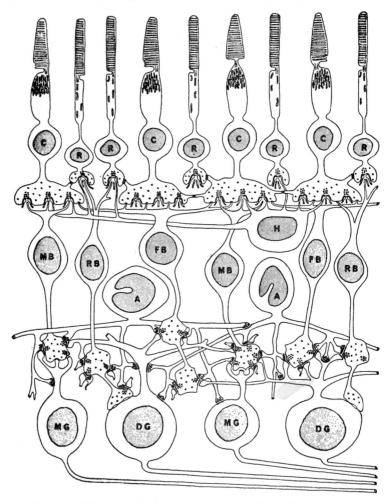

FIG. 3.1. Diagram of the synaptic connexions of the primate retina (rhesus monkey and man). R, rod; C, cone; MB, midget bipolar, probably receiving only from one cone; FB, flat bipolar, receiving from several cones, but probably from no rods; RB, rod bipolar, receiving from several rods, but probably from no cones; H, horizontal cell; A, amacrine cell; MG, midget ganglion cell; DG, diffuse ganglion cell (from Dowling & Boycott, 1966).

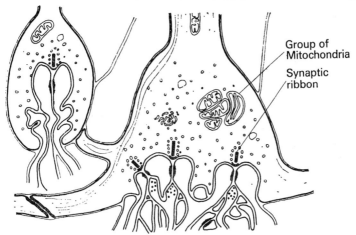

Group of
Mitochondria

Synaptic
ribbon

FIG. 3.2. Diagram of a rod spherule (left) and a cone pedicle (right)
from the human retina (from Missotten, Appelmans & Michiels,
1963).

ribbon and a cluster of synaptic vesicles. Rods also make synaptic
contact with processes of horizontal cells, but whether the same
horizontal cell receives synapses from cones and rods, as shown
in Fig. 3.1, is uncertain.

Besides the synapses with the triads, which by analogy with
similar structures elsewhere in the nervous system must almost
certainly work by chemical transmitters, the terminal pedicles of
cones make three kinds of junction at which clusters of synaptic
vesicles are absent. These are junctions with flat bipolars (FB),
with the pedicles of other cones, and with rod spherules. Such
junctions may be electrically transmitting synapses, or perhaps
not synapses at all.

Synapses conducting from horizontal cells to other cells have
not been firmly identified in primates. But in the cat and rabbit
Dowling, Brown & Major (1966) have found typical synapses,
with vesicles and membrane thickening, from horizontal cells to
bipolar cells.

The axons of most bipolar cells end in large terminal expansions
in the inner plexiform layer of the retina. Here they make synapses
of two kinds, both provided with synaptic vesicles and hence
presumably chemical in action. The first kind have synaptic
ribbons as well as vesicles, and transmit from the bipolar terminal

to two structures, a dendrite of a ganglion cell (DG or MG) and a process of an amacrine cell (A). The second kind appears to transmit in the opposite direction, i.e. from an amacrine cell process to the bipolar terminal. Figure 3.1 shows both kinds, in a relation which Dowling and Boycott think is common, perhaps universal: the same amacrine process that is influenced by the bipolar terminal through one synapse, influences it through another.

A few bipolar cells, probably only rod bipolars, have long axons that end in contact with the cell bodies of diffuse ganglion cells (DG). At these junctions there are no synaptic ribbons, clusters of vesicles, or membrane thickenings, but there are places where the two cell membranes are in direct apposition ('tight junctions'); it is suggested that these may be an electrically transmitting synapses.

There are also synapses, of presumably chemically transmitting kinds, between amacrine cell processes, and from amacrine cell processes to ganglion cells.

## Histological identification of cells that have recently been active

Nucleic acids (both DNA and RNA) are the only abundant constituents of cells that absorb heavily at wavelengths around 265 nm. Thus by photography or spectrophotometry at high magnification at these wavelengths, the amount of nucleic acid in single cells in ordinary histological sections can be roughly estimated (Brattgård, 1951). Brodskiǐ & Nechaeva (1958) and Utina, Nechaeva & Brodskiǐ (1960) showed that if frogs are kept in conditions where retinal ganglion cells are known to be very active (flickering light), the amount of nucleic acid in these cells rises; if the frogs are kept in conditions where retinal ganglion cells are known to be inactive or nearly so (darkness or steady light) the amount of nucleic acid falls. Half an hour of exposure to a given state of illumination suffices to give a clear change in the ultraviolet absorption of the cells. The changes are presumably in RNA, not DNA. It is unproved whether they are in the nuclei, the perinuclear cytoplasm, or both; but this doubt does not affect the value of the changes as an index of activity. As such an index, they have been used by Utina (1960) to examine the sensitivity of the bipolar cells of the frog to various frequencies of flicker. She found that light flickering at 0·5 c/sec raises the nucleic acid content of all bipolar cells, but light flickering at 4 c/sec raises only that of the outer layer of bipolar cells; this distinction fits well with inferences made by Byzov (1959) from electrical recordings.

The same technique has been used by Utina (1964) and Utina & Byzov (1965) to examine the flicker sensitivity of horizontal cells, amacrine cells and rods and cones. It could presumably be used to examine which cells become active when a patterned stimulus is presented, and this might be very informative.

Enoch (1966) has found that the staining properties of the ellipsoids of rods are affected by the amount of light to which the eye has been exposed during the 5 minutes before the rat is killed. The change in staining is presumed by Enoch to depend only on the amount of rhodopsin bleached. If so, it is unlike the change in ultraviolet absorption examined by Utina & Byzov (1965), which is much greater for flickering light than for the same amount of light put in steadily.

## The electroretinogram (e.r.g.)

As long ago as 1865 it was discovered by Holmgren that the difference of electrical potential, already known since the work of du Bois-Reymond (1848) to exist between the cornea and the back of the eye, altered when light fell upon the retina. This electrical response to illumination is now generally known as the electroretinogram. Kühne & Steiner (1880) showed that the retina itself is the source, by removing the retina from the opened excised eye of a frog, and finding that its electrical responses to light were similar to those of the whole eye. The shell of sclera, choroid and pigment epithelium left behind after removal of the retina was entirely unresponsive.

The time-course of the electroretinogram became accurately known almost as soon as sufficiently rapidly responding sensitive instruments were available to record it. Gotch (1903) determined its main features with the capillary electrometer, and von Brücke & Garten (1907) and Einthoven & Jolly (1908) established its details by means of the string galvanometer. One of Einthoven and Jolly's records is reproduced in Fig. 3.3.

*The similarity of the electroretinograms of different vertebrates*

The comparative studies of von Brücke & Garten (1907), Piper

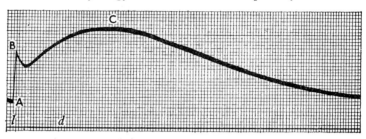

FIG. 3.3. Response of the unopened excised eye of a dark-adapted frog to a flash of light of duration 10 msec. Increasing positivity of the corneal electrode is shown upwards. The interval between adjacent vertical lines represents 0·5 sec, and that between adjacent horizontal lines 10 μV. If a stimuli of longer duration is given, the a-, b- and c-waves are not much altered, but a cornea-positive d-wave appears superimposed on the c-wave immediately after the end of the stimulus (from Einthoven & Jolly, 1908).

(1911) and later investigators have shown that the main features of the electroretinogram are strikingly similar throughout all the classes of vertebrates. Illumination of the eye causes a cornea-negative a-wave, and then a cornea-positive b-wave. These are differently affected by the brightness of the light; for strong stimuli they may be about equal in amplitude, but weak stimuli can be found that produce large b-waves but no detectable a-waves.

In the dark-adapted but not in the light-adapted eye the a- and b-waves are followed by a much slower c-wave, which is usually cornea-positive, but in the dog and rat (Parry, Tansley & Thomson, 1953) can be cornea-negative.

When the light is extinguished, in cold-blooded vertebrates, birds and a minority of mammals a cornea-positive response similar in its time-course to the b-wave is produced. This 'off-effect' or 'd-wave' is generally more prominent in the light-adapted than in the dark-adapted eye. The majority of mammals, including man, differ from other vertebrates in that the d-wave is very small, and is sometimes cornea-negative.

### Analysis into components

During the first third of the present century, provoked by the appearance of complexity in the electroretinogram and by the fact that different features of it were differently affected by such factors as dark-adaptation, at least seven attempts were made to identify a number of components, presumed to be of different origin, the sum of which constituted the response recorded from the whole eye. The most influential and best substantiated analysis is the latest, that of Granit (1933), which recognizes three components, called PI, PII and PIII (see Fig. 3.4). The analysis was based on

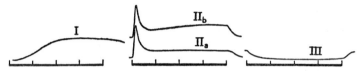

FIG. 3.4. Granit's analysis of the cat's e.r.g. for a stimulus lasting 2 seconds. IIa and IIb are alternative estimates of the course of the component PII. Time-marks 0·5 sec (from Granit, 1933).

the effects of ether anaesthesia and of asphyxia. The changes in the electrical record during deepening anaesthesia were interpreted as due to the removal first of PI and then of PII, and those produced by asphyxia as due to selective removal of PII.

The distinctness of PI (the c-wave) from the rest of the electroretinogram is now established beyond doubt by the results of recording with microelectrodes, and recently a response with the time-course of PIII has been similarly isolated (see p. 69). It is not, however, likely that the PIII that can be isolated with microelectrodes is identical with the PIII that is revealed by chemical agents, or that PIII is simple in origin.

Until 1950 (when intraretinal recording was introduced), the available evidence about the origin of the components of the electroretinogram was indirect. The observation of Granit and Helme (1939) that antidromic electrical stimuli to the optic nerve have no influence on the form of the electroretinogram showed that the ganglion cells make no substantial contribution, but did not help to decide whether the rods and cones, bipolar cells, horizontal cells, amacrine cells, or all of these are the generators. One apparently good piece of early evidence proved to be misleading: Adrian & Matthews (1927) found, for the frog and conger eel, that the latency of the electroretinogram lengthened as the area of retina illuminated decreased, and Granit (1933) found the same for the cat. The influence of area on latency might be taken as evidence that the earliest part of the electroretinogram is not generated by the rods and cones, for presumably a rod or cone cannot during the latent period be influenced by whether or not its neighbours are being illuminated. However, it is now clear that in these early experiments the responses recorded were not mainly due to light falling on the area of retina intentionally illuminated, but to stray light reflected or scattered on to other areas. The intensity of this stray light would be approximately proportional to the area of the test field, and the apparent effect of area on latency merely a result of the well-known effect of intensity on latency. That stray light was important in this way was elegantly shown when Boynton and Riggs (1951) and Asher (1951) found that in man a flash subtending a small angle at the eye gave a slightly larger electroretinogram when it fell on the blind spot than when it fell on retina. If precautions are taken against stray light, it is possible to obtain an electroretinogram from quite a small area of retina (Brindley, 1956b, 1957b, Brindley & Westheimer, 1965). The response then obtained, for a constant test illumination, is simply proportional to the area of the image of the test field on the retina, the latency being unaffected, as is illustrated in Fig. 3.5. There is no interaction between one area of retina and another in generating the electroretinogram, even when the regions illuminated are small and near to each other. This, as we shall see, contrasts with a conspicuous spatial interaction shown by some of the kinds of slow electrical activity that can be recorded with intraretinal microelectrodes.

*Examination by means of intraretinal extracellular microelectrodes*

The first experimenter to record electrical responses to illumination from within the retina was Tomita (1950), who removed the cornea, iris and lens from frogs' eyes and inserted a saline-filled glass capillary microelectrode into the retina from its inner surface. He found that when the tip of the microelectrode was at the inner surface of the retina, a typical normal electroretinogram was recorded between it and a large electrode in contact with the sclera. As the microelectrode was advanced through the retina, the form of the record changed: at depths between 70 and 210 μm from the inner surface, the polarity of both on- and off-effects was reversed, the microelectrode becoming negative in relation to the sclera at times corresponding approximately to the b- and d-

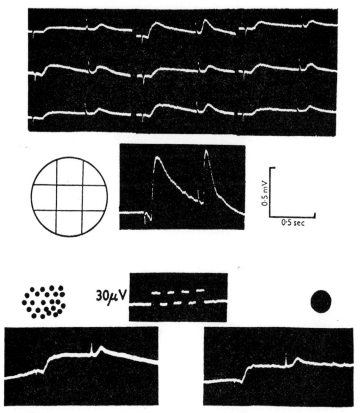

FIG. 3.5. Proportionality of the e.r.g. to the amount of retina illuminated. *Above:* responses to illumination of the whole eye and of nine parts of it (shown in the diagram on the left) separately. The large response is approximately the sum of the nine small ones. Stray light falling on parts of the retina outside the optical image of the stop determining the stimulus is prevented from causing any electrical response by superimposing the stimulus on a steady background of one-tenth of its intensity. *Below:* closely similar responses are obtained by illuminating a given total area of retina whether it is made up of twenty-four small fields (left) or one large one (right). The e.r.g. depends on the area of the region illuminated, and not on the length of its boundary. Time-scales are the same in upper and lower parts of the figure; the voltage-scales differ as shown in the calibrations. In this figure and in Fig. 3.6, the beginning of the stimulus is indicated by a downwardly directed artifact and its end by an upwardly directed one (from Brindley, 1956*b* and 1957*b*).

waves of the e.r.g. Advancing the electrode further caused the response to disappear or greatly diminish.

Similar experiments by Brindley (1956c, 1958) on frogs and by Brown & Wiesel (1961) on cats have given broadly similar results. An example is shown in Fig. 3.6.

The sudden great diminution of the response between the last two records of Fig. 3.6 occurs at an average depth of about 230 µm in the frog and 200 µm in the cat. It is reversible on withdrawing the electrode, and corresponds to penetration by the electrode of a barrier of high electrical resistance and capacity. When first discovered (Brindley, 1956a) this barrier was given the non-committal name 'R membrane', and tentatively identified with the external limiting membrane and the surface membranes of the rods and cones. Brown & Wiesel (1959) argued against this identification, and in favour of the view that the R membrane is Bruch's membrane. But it is now clear that both the early identifications were wrong. The R membrane has two parts. The principal one is the inner bounding membrane of the pigment epithelium, as Brindley & Hamasaki (1963) argued from electrical evidence, Cohen (1965) supported by electron-microscopical evidence, and Rodriguez-Peralta (1968) confirmed by examining permeability to dyes. This pigment-epithelial component has all the electrical properties of the whole R membrane; that is, if a steady current $i$ is suddenly passed through either the whole R membrane or its pigment-epithelial component, the potential developed across it is described by $E = ia(1 - e^{-t/\alpha}) + ib(1 - e^{-t/\beta}) + ic$, where $\alpha \simeq 10$ msec, $\beta \simeq 0.3$ msec, and $a$, $b$ and $c$ are arbitrary constants which differ little if at all between the whole R membrane and its pigment-epithelial component. However, Byzov and Hanitzsch (1964) and Byzov (1968) showed that in isolated pars optica retinae there is an electrical barrier with some but not all of the properties of the R membrane; it lacks the shorter of the two time-constants, and the resistances $a$ and $c$ are smaller for it than for the pigment-epithelial component. It is an open field for speculation whether the high electrical resistance (i.e. impermeability to ions) of the R membrane is functionally important. All that we now know is that it helps experimenters to know the position of microelectrodes.

The inversion of the response when the electrode is more than about 150 µm from the inner surface of the retina indicates that

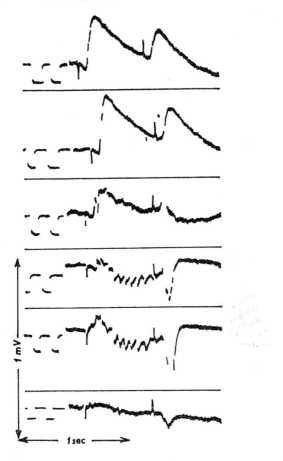

FIG. 3.6. Responses to uniform illumination of the whole retina of a freshly excised eye of *Rana temporaria*, recorded with an intra-retinal microelectrode. From above downwards, the electrode was at 5, 55, 105, 155, 205, and 255 $\mu$m respectively from the inner surface of the retina. This pattern of responses was first described by Tomita (1950). On the left are shown, for the same electrode positions, the potential pulses recorded when square pulses of current are passed between vitreous and sclera (from Brindley, 1957*a*).

it has then passed through the layer of dipoles that generates the electroretinogram. It is inverted rather than lost because, owing to conduction round the edges of the opened eye, the potential of the 'indifferent' electrode in contact with the sclera is, roughly speaking, a weighted mean of the potentials of the inner and outer surfaces of the retina.

If the sole electrically active structures were dipoles that formed a single uniform layer, then all parts of the intraretinally recorded response should diminish, disappear, and then invert at the same time, so that a record from any depth in the retina should be convertible into that from any other depth by multiplication by a constant factor, positive or negative. If there were two such uniform layers of dipoles overlapping by a constant amount, then on a plausible set of assumptions about the passive electrical properties of the system one might expect the record obtained from any depth to be derivable as a linear function of two fixed curves, one corresponding to each layer of dipoles. As can be seen in Fig. 3.6, and in many other published records, the true situation is more complex than this; and if one tries to construct pairs of fixed curves from which the records can be *approximately* derived as linear functions, one finds that neither can be made to resemble the PIII that can easily be obtained in the frog by the use of chemical agents (e.g. Hamasaki, 1964). The same holds for Tomita's records (1950, 1963). In the cat (Brown & Wiesel, 1961) the intraretinally recorded responses seem to be equally complex, though the complexities are of a different kind.

The conclusions that we can draw about the origin of the electroretinogram from the results of intraretinal recording are far from solving all its problems. The layer of ganglion cells contributes nothing, for nothing can be recorded differentially across it. There are more than two kinds of generator of slow electrical activity, and they overlap spatially. It cannot necessarily be concluded that a generator that contributes to the intraretinal records must contribute to the electroretinogram; it may be a tangentially orientated dipole or a radially symmetrical tripole. In the cat, the records of Brown and Wiesel show fairly clearly that part at least of the a-wave arises in a layer external to that which is mainly responsible for the b-wave. In the frog, such a spatial separation of the sources of the a- and b-waves is more difficult to detect, but there is an indication of it in some of the older published records, including that reproduced in Fig. 3.6.

A really clear separation by microelectrodes of a component of the electroretinogram was first achieved by Murakami & Kaneko (1966). They placed a retina of a frog, carp or turtle on a metal plate with the receptors uppermost, and recorded responses with a coaxial double microelectrode as shown in Fig. 3.7a. When the inner electrode was at a depth of 120 μm, an electroretinogram of nearly normal form could be recorded between the inner electrode and the vitreal surface of the retina, and a response resembling PIII between the inner and outer electrodes (Fig. 3.7b). Murakami and Kaneko call the upper response of Fig. 3.7b 'distal PIII' because they believe, on good indirect evidence, that there is another component of the electroretinogram, very similar in time-course but arising more proximally; this they call 'proximal PIII'. They suggest that distal PIII is generated by rods and cones, and the suggestion gains strong support from the intracellular recordings of Tomita (1965) and Kaneko & Hashimoto (1967) that are discussed in Chapter 2.

There are probably two reasons why Murakami and Kaneko succeeded in clearly separating a PIII-like component with microelectrodes where many earlier experimenters had tried and failed. The lesser reason is that direct recording (with a coaxial double electrode) of the difference of potential between two neighbouring layers eliminates accidental disturbances better than does inferring it by subtracting two records, though ideally both methods should give the same answer. The greater reason for the success of Murakami and Kaneko must be that their removal of the pigment epithelium allowed their outer microelectrodes to be brought in contact with the outer surface of the receptors without the intervention of the highly resistive pigment-epithelial part of the R membrane; in microelectrode experiments on intact eyes or opened excised eyes the outer surface of the receptors is never directly accessible to the electrode.

Brown & Watanabe (1962) provided two arguments in favour of a receptoral origin for part, at least, of PIII. The first was that occlusion of the retinal circulation in monkeys, the choroidal circulation remaining intact, isolates PIII. It was supposed that the choroidal circulation supplies only the receptors and that none of the nerve cells could function without the retinal circulation. This argument was weak, as became clear when Fujino &

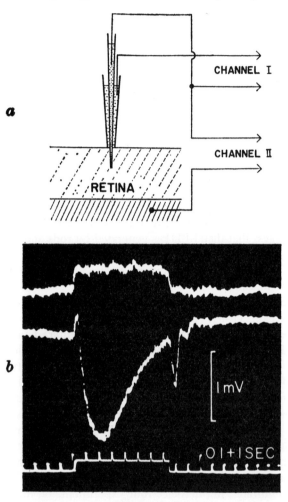

FIG. 3.7. (*a*) Coaxial double microelectrode for recording differentially between a point within the retina and a point on the surface very close to it. (*b*) Responses recorded differentially between the intraretinal pipette and the receptoral surface (upper record) and the vitreal surface (lower record) of the retina (from Murakami & Kaneko, 1966).

Hamasaki (1965) showed that occlusion of the choroidal circulation, the retinal remaining intact, also isolates PIII. The second argument was stronger: if local responses from foveal and peripheral retina of a diurnal monkey are compared, those from the fovea (where receptors are abundant but there are no nerve cells near) have a larger a-wave and a much smaller b-wave than those from peripheral retina.

It is of some importance to know whether the 'late receptor potential' (Brown and Watanabe's name for the PIII-like response isolated by their technique of occluding the retinal circulation) is mainly of receptoral origin, because it has been used, notably by Maffei & Poppele (1968), to analyse dark-adaptation and the processing of temporal patterns by the retina. Almost certainly part of it comes from receptors and part (as we shall see later) from horizontal cells; but the relative size of these two constituents is difficult to determine.

The largest contributors to the electroretinogram, especially to its b-wave, are the bipolar cells, as was probable already by 1959 (see the first edition of this book), but was first clearly proved by Hashimoto, Murakami & Tomita (1961). They recorded responses from frogs' retinas by means of a coaxial double electrode like that of Fig. 3.7a. At sites where maximal responses were recorded, they expelled ferricyanide ions electrophoretically from the inner pipette to mark (after reaction with externally applied ferrous chloride) the site of recording. Histological examination showed that the largest responses were found when the tips of the two pipettes (kept at a constant distance of 50 μm) were on opposite sides of the layer of bipolar cell bodies.

## The origin of the c-wave

The early investigators from Einthoven and Jolly (1908) to Granit (1933), all agreed in attributing to the c-wave (PI) an origin different from that of the rest of the electroretinogram. Riggs & Johnson (1949) and Dodt (1951) found that the human c-wave is sometimes abolished by atropinization, and it is clear that pupillary reflexes can contribute substantially to what is, descriptively, a c-wave. However, the c-wave is certainly not wholly of pupillary origin, for it has been recorded from opened excised frogs' eyes with the iris cut away (Piper 1911), from cats and dogs in which the pathways of the pupillary reflex were wholly interrupted (Granit, 1933; Parry, Tansley & Thomson, 1953) and from freshly excised eyes of monkeys (Fujino & Hamasaki, 1965).

Noell (1954) suggested that the non-pupillary part of the electroretinogram might be generated by the pigment epithelium, principally on the ground that when rabbits are poisoned with iodate, which severely injures the pigment epithelium without causing any histologically detectable injury to the retina, the c-wave is abolished and the a- and b-waves unaffected. In the cat, Brown & Wiesel (1958, 1961) recorded the c-wave with intra-retinal microelectrodes, and found that it reverses polarity when the electrode penetrates the R membrane. They accepted this finding as confirmation of Noell's view. But if the pigment epithelium generates the c-wave, it must be the rods and cones, or the rods alone, that provoke it to do so, for the spectral sensitivity of the c-wave corresponds to the absorption spectrum of rhodopsin, not to that of melanin (Granit & Munsterhjelm, 1937; Dodt, 1957). And it is very unlikely that the pigment epithelium is the sole source, for Sickel (1966) was able to record c-waves from the isolated pars optica retinae of the rabbit. The reversal of the polarity of the c-wave across the R membrane is compatible with its being generated by both the components (receptoral and pigment-epithelial) of that membrane, or even by the receptoral component only.

Whatever its source, the c-wave is evidently no part or direct consequence of the message transmitted to the brain by the retina; it is far too slow.

### Rod and cone electroretinograms

In mixed retinas, such as those of man and the frog, electro-retinograms can be obtained in dark-adapted and light-adapted states, the spectral sensitivity changing with adaptation from that characteristic of rods to that characteristic of cones. The responses in the two states are alike in their main features, but differ in details: in the light-adapted eye the c-wave is absent, the off-effect is larger in proportion to the on-effect, and the time-course of every part of the response is shorter.

Such differences in form between the dark-adapted and light-adapted electroretinograms do not necessarily represent a change from rods to cones; for example, the electroretinogram of the guinea pig is much affected by the state of adaptation (Boehm, Sigg &

Monnier, 1944), though its retina contains practically no cones; and the records of Riggs, Berry & Wayner (1949) show that, in the human eye, degrees of light-adaptation that are insufficient to convert the spectral sensitivity curve from the scotopic to the photopic nevertheless suffice to alter the shape of the b-wave.

The proper technique for looking for differences between the electroretinograms provoked by activity of rods and of cones is to compare the responses of a mixed retina in a constant adaptational state to lights of different wavelength or, even better (see p. 249), different directions of incidence on the retina. In the frog, even in the most favourably intermediate states of adaptation, the responses to red light, stimulating mainly cones, and to blue light, stimulating mainly rods, are almost the same. Minute differences have been described by Kohlrausch (1918) and by Goto & Toida (1954), but by far the greater part of the large difference between the frog's photopic and scotopic electroretinograms is unrelated to the replacement of rods by cones as light-adaptation proceeds. This need not be thought very surprising, since rods and cones in the frog largely though not wholly share the same nervous pathways. In man we must suppose (in accord with electron-microscopical evidence) that there is less sharing of bipolar cells between rods and cones, for conspicuous effects of wavelength on the electroretinogram can be demonstrated in a single adaptational state (Adrian, 1945; see Fig. 3.8). For the light-adapted eye the responses produced by different colours are all brief and diphasic, a very small cornea-negative deflexion (a-wave) being followed by a larger cornea-positive one (b-wave). The response to deep red light is the same in the dark-adapted eye as in the light-adapted, but with blue and green stimuli the brief diphasic response given by the light-adapted eye is replaced or masked, in the dark-adapted state, by a slower and much larger cornea-positive deflexion. Orange-red and yellow flashes give, in the dark-adapted eye, two-humped responses such as might be expected if the types of responses given by blue and deep red light were being provoked simultaneously. Evidently a slow cornea-positive deflexion is produced when rods are stimulated, and a faster diphasic response, the first phase negative, when cones are stimulated. Later experiments have confirmed this distinction, but introduced some modifications; in particular, the human a-wave is not purely of cone origin. In the response of the

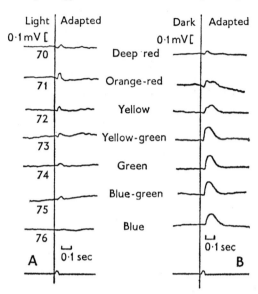

FIG. 3.8. Electrical responses of the human eye to brief flashes of
light of various colours (from Adrian, 1945).

dark-adapted eye to bright flashes, Armington, Johnson & Riggs
(1952) found that it often consists of two clearly distinguishable
troughs. The first of these has roughly the spectral sensitivity of
photopic vision and is very little affected by light-adaptation, like
the a-wave of Adrian's records. The second is much depressed
by light-adaptation, and has nearly the spectral sensitivity curve
of scotopic vision, so that it almost certainly depends on the
stimulation of rods.

### Miscellaneous properties of the electroretinogram

It is easy to do exploratory experiments on the electroretinogram, and such experiments often yield results that are clear and
not without interest, but do not, in the present state of knowledge,
enlarge our understanding of how the retina works. On page 68
of the first edition of this book many such results were briefly
mentioned. It would be easy now to reproduce this page (for it has
neither gained nor lost significance), and to add another hundred
references to cover the last nine years; but I will restrict myself

to giving one fact. I cannot interpret it but have a strong intuitive feeling that it will become significant when more is known.

Cone & Platt (1964) examined in the rat the electroretinograms produced by stimuli superimposed on steady backgrounds. They found (as holds also in the frog: Brindley, 1956b) that the amplitude and general shape of the responses is determined, over a wide range, by the ratio of the intensity of the incremental stimulus to that of the background. The new discovery of Cone and Platt was that responses for the same ratio of intensities but different absolute values of intensity are displaced on the time axis: the latency, and in general the time from the stimulus to any given feature of the response, are shortened by any increase of absolute intensity at constant ratio. This effect of timing can be seen also in my old published records from frogs, but less conspicuously, and I failed to notice it.

## The responses of single nerve cells of the retina

When *nerve impulses* (spike potentials) are recorded by means of a microelectrode from within a piece of nervous tissue, it is often possible to be confident that they come from a single cell, even though the tip of the electrode is in extracellular space; the large impulses that come from the cell that is nearest to the electrode may be easily distinguishable from the small ones that come from more distant cells. Such extracellular recording has been extensively used to study the impulses produced by ganglion cells since 1939. It is possible also to use it to study impulses produced by certain cells of the inner nuclear layer (Brindley, 1956c; Brown & Wiesel, 1959; Kaneko & Hashimoto, 1968), but the tight packing of these cells makes the recognition of single units difficult.

When we wish to study *graded slow electrical responses*, such as are produced by many cells of the retina, there is no way of isolating the response of a single cell except by placing the tip of the electrode within it. This is technically difficult because the cells of the retina are small, and it is only recently (Kaneko & Hashimoto, 1967; Werblin & Dowling, 1969) that it has been *proved* to have been done by injecting dye from the tip of the electrode and showing by subsequent histological examination that the dye is restricted to the interior of one cell. However, by using this new knowledge we can infer that some records obtained long ago (e.g. Svaetichin, 1953) were very probably intracellular.

### Rods and cones

The electrical responses that can be recorded from within cones (and perhaps also rods) were considered in Chapter 2. In Fig. 3.9 they are compared with responses of other cells of the retina.

### Horizontal and bipolar cells

Svaetichin (1953) discovered the slow electrical responses that are now generally known as s-potentials. They can be recorded in mammals, reptiles and amphibia (for the cat see Grüsser, 1957 and Brown & Wiesel, 1958) but have been most thoroughly studied in fishes. A microelectrode slowly advanced to the appropriate depth in the retina suddenly records a negative resting potential, usually between 20 and 50 mV, and as long as this resting potential is present illumination of the retina causes a response, graded with the strength of the stimulus, that can be as large as 30 mV. These s-potentials are of two kinds, colour-sensitive and non-colour-sensitive. The non-colour-sensitive (often called L-type) s-potentials are always negative-going, i.e. correspond to a hyper-polarization of the bounding membrane of the space (whether it be a cell or not) from which they are recorded. Though non-colour-sensitive in this sense, they are not excited only by means of a single kind of receptor. Naka & Rushton (1966b) showed that in the tench (*Tinca tinca*) they receive messages from at least three kinds of receptors, presumably cones, with spectral maxima of sensitivity probably near 540, 620 and 680 nm. The colour-sensitive s-potentials are negative-going for some wavelengths of light and positive-going for others. The time-courses of the two kinds of responses differ slightly. Both are simpler than the electroretinogram, and, as we shall see, both probably contribute to it. It was shown by MacNichol & Svaetichin (1958) that the sites from which both kinds are recorded are in the inner nuclear layer, the non-colour-sensitive in the region of the horizontal cells and the colour-sensitive slightly nearer to the vitreous, in a region that contains both horizontal and bipolar cells. The space from which a given response (of non-colour-sensitive type in fish) can be recorded is so large that it is not easy to reconcile it with a single cell (Tomita, 1957); nevertheless, on the balance of

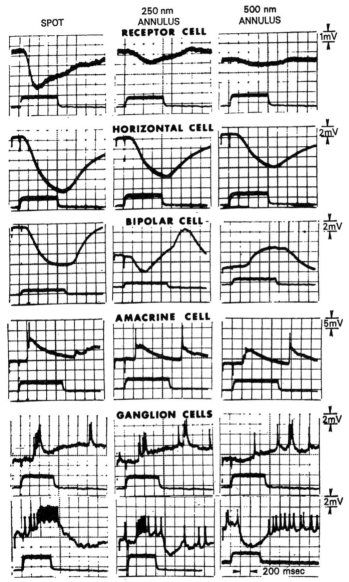

FIG. 3.9. Responses recorded intracellularly from cells of the retina of *Necturus*, an amphibian that has the advantage for this purpose that its cells are exceptionally large. Spikes are seen only in records from amacrine cells and ganglion cells. These spikes appear on the records at far less than their true amplitudes (presumably about 100 mV) because of attenuation of high frequencies by the electrode (from Werblin & Dowling, 1969).

rather weak conflicting evidence many investigators have thought it likely that all s-potential spaces are cells. If so, the non-colour-sensitive responses must come from horizontal cells. The colour-sensitive might come from either horizontal or bipolar cells. The alternative view, supported by Tomita's demonstration of their large size, would be that the s-potential spaces are regions of extracellular space bounded by tightly joined neurons or glia cells or both. The balance of probability between these two interpretations has been a good deal altered by the experiments of Werblin & Dowling (1969), who have proved clearly by marking with dye that in the amphibian *Necturus*, whose retinal nerve cells are exceptionally large, responses very similar to the s-potentials of fishes are recorded intracellularly from horizontal and bipolar cells. The responses of the horizontal cells of *Necturus* are very similar to the non-colour-sensitive s-potentials of fishes. The responses of the bipolar cells of *Necturus* resemble in some respects the colour-sensitive s-potentials of fishes, for they are negative-going for some stimuli and positive-going for others. But the property of the stimulus that determines the polarity of the response is not colour, but position. The responses are negative-going (hyperpolarizing) for stimuli in the centre of the receptive field of the bipolar cell and positive-going (depolarizing) for stimuli in the periphery of the field (see Fig. 3.9), provided that the cell is somewhat hyperpolarized when the peripheral stimulus is given. The colour-sensitive s-potentials of fish, at least in *Cyprinus auratus*, do not share this spatial property (Watanabe & Tosaka, 1959).

Though, in time-course, s-potentials roughly resemble post-synaptic potentials as seen, for example, in motor neurons, the mechanism of their generation seems to be quite different, for the membrane resistance of the spaces from which they are recorded alters very little (Gouras, 1960; Tomita & Kaneko, 1965) or not at all (Tasaki, 1960; Watanabe, Tosaka & Yokota, 1960) during the responses.

The relation of s-potentials (of the non-colour-sensitive kind) to the intensity and spatial distribution of the stimulus has been examined by Naka & Rushton (1967) and found to be unlike that of any other known kind of single-cell response in the retina, though it is similar to that of the electroretinogram (see p. 64 and Fig. 3.5); that is, the response to spatially separate or very differently coloured (red and blue) stimuli add linearly, though those to stimuli that are spatially coincident and similarly coloured add very non-linearly in the sense that the response to two stimuli together is much less than the sum of the responses to them separately.

Non-colour-sensitive s-potentials share some at least of those properties of PIII of the electroretinogram that Cone & Ebrey (1965) pointed out as objections to the view that PIII is an essential part of the message transmitted by the retina. Thus Witkovsky (1967) found that the thresholds of non-colour-sensitive s-potentials were 100 to 1000 times higher than those of ganglion cells, and that s-potential sources were far less responsive than ganglion cells to incremental stimuli. Whether or not these are valid objections to the view that s-potentials are an essential part of the retinal message (and on the whole I think they are not), they certainly form part of the close general similarity between non-colour-sensitive s-potentials and PIII. This similarity strongly suggests that horizontal cells generate the component of the electroretinogram called 'proximal PIII' by Murakami & Kaneko (1966), though it is difficult to test by direct experiment whether they do so.

Ostrovskii & Trifonov (1967) found that mechanical damage to the receptors in the neighbourhood of a generator of s-potentials decreases the response of the generator to light, as one would expect, but increases the resting potential. From this and related evidence they make the interesting suggestion that the chemical transmitter from receptors to horizontal cells causes depolarization of the horizontal cell, that in darkness this transmitter is continuously released, and that illumination of the receptors blocks its release. This hypothesis is the most economical explanation of their findings, but the opposite view that the transmitter hyperpolarizes horizontal cells, and light and mechanical damage both provoke its release, has not been excluded.

Trifonov (1968) and Byzov & Trifonov (1968) found in terrapins (*Emys orbicularis*) and carp (*Cyprinus carpio*) that s-potential sources give depolarizing responses to brief electrical stimuli put through the whole retina in the sclera-positive direction. The responses last about 30 msec, saturate with large inputs, and have a threshold. Stimuli of opposite polarity are either ineffective, or also give depolarizing responses, but smaller ones. The responses to electricity and to light interact in a rather complex manner. Trifonov (1968) has also examined the effects of prolonged steady currents passed through the retina upon the generators of s-potentials. Such currents, unlike brief pulses, have approximately symmetrical opposite effects for opposite polarities. Trifonov argues, very plausibly though not absolutely compellingly, that in all the phenomena that he studies the current is acting on the presynaptic membrane of receptor-horizontal cell synapses.

Kaneko & Hashimoto (1969) have recorded intracellularly from cells of the inner nuclear layer in carp. The cells recorded from were certainly either bipolar or amacrine cells but it is not possible to decide which. Some cells gave depolarizing and some hyper-polarizing responses. Many gave responses whose polarity depended on the position of the stimulus in their receptive field, as Werblin and Dowling found in bipolar cells of *Necturus*, and for a small minority the polarity of response depended on the colour of the light. Upon depolarizing responses, spikes were superimposed in some cells, and regular oscillations of potential in others.

*Amacrine cells*

The responses of these are known clearly only from the work of Werblin and Dowling on *Necturus* (Fig. 3.9). Any stimulus that affects them at all depolarizes them. When the depolarization reaches a critical level, spikes are produced. The spatial distribution of the stimulus determines whether the amacrine cell responds to the onset of illumination, to its cessation, or both.

Amacrine cells are the only cells of the inner nuclear layer from which Werblin and Dowling were able to record spikes. This suggests that the cell of the inner nuclear layer from which Brindley (1956c) and Brown & Wiesel (1959) obtained spikes by extracellular recording in frogs and cats may have been amacrine cells. The units studied by Brown and Wiesel in cats had a spatial organization of sensitivity very similar to that described by Werblin and Dowling in *Necturus*.

Kaneko & Hashimoto (1968) have shown that the spike-producing units of the inner nuclear layer of the frog are not merely displaced ganglion cells, since ganglion cells can be excited by antidromic stimulation of the optic nerve, but the spike-producing units of the inner nuclear layer cannot.

*Ganglion cells and optic nerve fibres*

Action potentials from the optic nerve were first recorded by Adrian & Matthews (1927, 1928*a*, *b*) using eyes excised, with the entire optic nerve attached, from pithed conger eels (*Conger vulgaris*). The records are those of many fibres together, but they show well the general pattern of activity. In darkness, the nerve is almost completely inactive. This inactivity may be a property of cold-blooded vertebrates as opposed to warm-blooded, or, less probably, of excised eyes as opposed to those with their circulation intact; the ganglion cells of decerebrate or anaesthetized cats generate frequent irregularly spaced impulses in the dark (Kuffler, FitzHugh & Barlow, 1957; Bornschein, 1958). The important but difficult question whether the spontaneous discharge in mammals is due to noise in the rods and cones (for example thermal decomposition of visual pigment molecules) is discussed by Hughes & Maffei (1965), Burke & Hayhow (1968) and Barlow & Levick (1969*a*). The arguments on neither side are compelling, but the view of Burke and Hayhow and of Barlow and Levick that the spontaneous discharge is of receptoral origin is more attractive on grounds of biological economy than the opposite opinion of Hughes and Maffei.

On illuminating the conger eel's eye, the previously inactive optic nerve gives, after a latent period of about 0·2 sec, a burst of impulses at high frequency during the first second, followed by a nearly steady discharge at much lower frequency lasting for as long as the eye is illuminated. When the light is extinguished, the frequency diminishes slightly for a few tenths of a second, increases substantially for a few more tenths of a second (off-discharge), and then ceases.

The contribution of individual nerve fibres, and therefore presumably of individual retinal ganglion cells, to the general pattern of activity of the optic nerve was analysed by Hartline (1938), who solved the technically very difficult problem of dissecting out small bundles of fibres on the surface of the frog's retina and cutting them down until only a single fibre remained active in each bundle. He found that fibres were of three kinds: about 20% gave a burst of impulses when the light was switched on, followed by a sustained discharge lasting for as long as illumination continued (on-fibres), 50% gave bursts of impulses when the light was switched on and when it was extinguished, but were inactive in steady light (on-off-fibres), and 30% responded only to extinction of the light (off-fibres).

Hartline's technique is too difficult to have been much imitated. Most later investigators have preferred the simple device introduced by Granit & Svaetichin (1939) of placing a microelectrode on the inner surface of the retina. It was at first uncertain whether the spikes recorded by such electrodes were derived from ganglion cells or optic nerve fibres, and whether the cells or fibres were typical or in some way exceptional, but Rushton (1949) showed clearly that in the cat the large well-isolated spikes to which the attention of investigators has been mainly directed are produced by the largest of the retinal ganglion cells. In many species the ganglion cells do not vary in size as widely as they do in the cat, and the isolation of single units in the excised frog's eye, for example, seems to depend upon which of the ganglion cells longest survive deprivation of blood, rather than upon their size.

Besides the spikes produced by ganglion cells those of optic nerve fibres running over the surface of the retina can sometimes be recorded with metal microelectrodes. They are smaller and less easy to isolate. They have the expected triphasic shape, in contrast to the diphasic spikes of the ganglion cells (Barlow, 1953*a*).

It is also possible to record from single optic nerve fibres by inserting a microelectrode into the optic nerve, chiasma or tract, and thereby avoid surgical interference with the eye.

*Effect of the intensity of the stimulus.* In contrast to the electroretinogram, but in agreement with much of the slow activity recorded with an intraretinal microelectrode, the responses of single ganglion cells are affected in a complex and variable way by the intensity of the stimulus. In steady light, only a very small minority of cells (Barlow & Levick, 1969*b*) give a discharge whose frequency increases with increasing illumination over a wide range. Most on-centre cells (see p. 87) show this kind of simple relation at very low illuminations, but they at higher illuminations and off-centre cells at all illuminations behave in a complex manner that varies from cell to cell.

Responses to flashes, steps and other transient stimuli, though they are often made larger by increasing the strength of the stimulus, are not always so (Hartline, 1938 and many others later);

and even when the relation of response to strength of stimulus is in the expected direction, it is rarely simple or regular.

*Responses to colour.* Granit & Svaetichin (1939), Granit (1941, 1943, 1947) and Donner (1953) investigated the threshold spectral sensitivities of single ganglion cells or small groups of ganglion cells in light-adapted and dark-adapted states in a wide range of vertebrates. In the dark-adapted state, the findings are relatively simple. In every species investigated that possesses a large proportion of rods in its retina, the spectral sensitivity curve of each ganglion cell corresponds to the absorption spectrum of the rod pigment; rhodopsin for the cat, rat, guinea-pig, rabbit, frog and eel (*Anguilla*), and porphyropsin for the tench (*Tinca vulgaris*). Small deviations from this rule have been found in several of these species, but for all of them it is at least a good approximation. When the eyes are light-adapted, this simple situation is replaced by great complexity. One fact is immediately clear: the spectral sensitivity curve of each ganglion cell changes, and cells which in the dark-adapted state had shown themselves to be connected to rhodopsin- or porphyropsin-containing rods now give evidence of being connected to other receptors instead or in addition. Often we may presume that these additional or substituting receptors are cones, but such a presumption is not always justified; the frog certainly has green rods besides those that contain rhodopsin, and there may well be other animals whose rods are of more than one kind.

The spectral sensitivities of ganglion cells in the light-adapted eye do not fall, for any one species, into a small number of clearly distinct classes, analogous to the three classes of cones which we believe, on the evidence of colour-matching experiments, to be present in the human eye. In the frog the commonest kind of ganglion cell is one with a broad spectral sensitivity curve whose maximum lies at about 560 nm. This type of cell was called by Granit the 'photopic dominator'. A closely similar spectral sensitivity curve is found in about one-third of the ganglion cells of the light-adapted cat (Granit, 1943) and in many of those of the pigeon, both in light-adapted and dark-adapted states (Donner, 1953). The majority of ganglion cells in the tench have a spectral sensitivity curve of similar shape, but with its maximum at 610 nm, so that the amount of the Purkinje shift is about the same for these cells in the tench (540 to 610 nm) as for the corresponding cells of the frog (505 to 560 nm).

The spectral sensitivity curves of ganglion cells that are not 'photopic dominators' are of very many kinds, even within one animal, and both they and the spectral sensitivity curves of the dominators themselves can often be greatly altered by adaptation of the eye to coloured lights. Granit suggested that the non-dominator ganglion cells, and possibly the dominators also, are built up from elements called 'modulators' with narrow spectral sensitivity curves; but this idea agrees poorly with modern knowledge of visual pigments if the modulators are supposed to be receptors, and is not supported by modern electrophysiology if the modulators are supposed to be horizontal or bipolar cells.

All the early quantitative work on the colour-sensitivities of single ganglion cells was done by measuring absolute thresholds for various wavelengths. The retina was usually not in adaptational equilibrium, and this made accurate measurement difficult. The sensitivity of a single ganglion cell is not a necessarily invariant property like that of a receptor containing only one pigment. There is no reason why it should be the same at incremental threshold, i.e. for stimuli added to a continuous uniform background, as at absolute threshold, or the same at threshold as at some constant non-threshold response. Evidence that the criterion of constant response can influence the spectral sensitivity curve appeared as early as Granit's paper of 1943 on the cat. Granit found that some ganglion cells consistently gave different patterns of response for different colours of light; for example, red light might cause brief bursts of impulses at 'on' and 'off' without other activity, but blue light a steady discharge during illumination in addition to the on- and off-responses. Other effects of colour on the pattern of discharge of single ganglion cells have been described by Donner (1949), Liberman (1957) and others. They make it clear that to state merely the threshold spectral sensitivity curve of a ganglion cell is to specify its properties in relation to wavelengths very inadequately, even for a single state of adaptation. Among the most striking effects are those described for the goldfish by Wagner, MacNichol & Wolbarsht (1960) and for the spider monkey by Hubel & Wiesel (1960). The majority of the ganglion cells of the goldfish and a small minority of the optic nerve fibres of the spider monkey give opposite responses to red and blue light: if one colour stimulates at 'on', the other stimulates at 'off'. An intermediate wavelength can be found that has almost no stimulating effect. The analogy with colour-sensitive s-potentials is obvious. Daw (1967) has shown that the colour-sensitive ganglion cells of the goldfish have a spatial antagonism between the central and peripheral parts of their receptive fields (see below), in addition to their chromatic antagonism. A comparable organization occurs in the ground squirrel (Michael, 1968).

Donner (1959) and Donner & Rushton (1959b) have described experiments on the thresholds of the frog's ganglion cells to incremental stimuli of one wavelength superimposed on a background of another wavelength. By this technique, which is closely analogous to one which Stiles has used extensively in human experiments (see pp. 231–235), a much more stable state of adaptation can be maintained than when measurements are made at absolute threshold, and spectral sensitivities can be measured much more accurately.

Another property of single ganglion cells which may help us to discover what kinds of receptors are connected to each cell and how they co-operate in exciting it is that certain substitutions of one stimulus for another can be made without causing any discharge of spikes by the cell. This is the analogue for a single ganglion cell of a colour-matching experiment for the whole human eye. Such experiments on ganglion cells of the frog have been described by Bongard & Smirnov (1957) and by Donner & Rushton (1959a). Bongard and Smirnov found that when a microelectrode was recording spikes which they believed came from a single ganglion cell, sudden replacement of any light by light of a different wavelength always produced a discharge, however the relative intensities were adjusted; but for light of any wavelength it was possible to find a mixture of red and blue which could be suddenly substituted for it without provoking any discharge. Thus according to Bongard and Smirnov a single ganglion cell of the frog is analogous to a dichromatic (protanopic, deuteranopic or tritanopic; see p. 227) human eye. Donner and Rushton found some cells which had the properties described by Bongard and Smirnov, but their investigations were mainly concerned with a surprising class of cells which could be proved to be functionally connected to at least two classes of receptors with different spectral sensitivities, but which nevertheless had properties analogous to

D

those of a monochromatic human eye, for example the dark-adapted eye in dim light. The sensitivity of these cells varied with the direction of incidence of the stimulating light upon the retina for long wavelengths (Stiles–Crawford effect; see p. 248), but did not for short wavelengths. Such directional selectivity can only be a property of receptors, and Donner and Rushton provided evidence (already available in Stile's work for human rods and cones) that a receptor that is directionally selective for one wavelength is directionally selective for all wavelengths that it detects. It is thus practically certain that the ganglion cells studied by Donner and Rushton received signals from two classes of receptors (very probably rods and cones) with different spectral sensitivities. It was found, however, that many of them could not detect sudden replacement of a light by another of different wavelength if the relative intensities of the two lights were appropriately adjusted, though they were sensitive to very small changes in intensity at any one wavelength. Evidently a decrease in the degree of stimulation of one class of receptors could compensate for a simultaneous increase in the degree of stimulation of another class.

*Spatial properties of ganglion cells.* Although in the human fovea histological evidence indicates that there are small ganglion cells each of which is connected only to a single cone, the usual situation throughout the vertebrates is for each ganglion cell to serve many receptors. The region of the retina within which these receptors lie is known as the receptive field of the ganglion cell. The organization of these receptive fields has been studied in detail in the frog, cat, monkey and rabbit.

The receptive field of a typical ganglion cell or nerve fibre in the frog (Hartline, 1940a, b; Barlow, 1953b) is roughly circular and between 0·5 and 1 mm in total diameter. A light falling near its margin has to be 1000 to 10,000 times brighter in order to stimulate than one falling on its centre. It has a central 'plateau', of roughly constant high sensitivity, whose diameter varies between 0·2 and 0·6 mm. The pattern of discharge is often independent of the position of the stimulus within the receptive field; but there are some 'on-off' ganglion cells which are about equally sensitive at 'on' and at 'off' to stimuli falling in the centre of the field, but have thresholds a hundred times lower at 'off' than at 'on' from stimuli falling near its margin. For pure 'off'

ganglion cells stimulated with spots of light of various sizes concentric with the receptive field, the sensitivity (i.e. the reciprocal
of the threshold illumination) is almost proportional to the area
provided that the stimulus does not extend beyond the central
plateau of the receptive field. Further increase in the area of the
stimulus produces a small further increase in the sensitivity. For
'on-off' ganglion cells, the sensitivity increases less than in proportion to the area of the stimulus as this is increased within the
bounds of the central plateau. When the diameter of the stimulus
exceeds about 0·8 mm, the sensitivity actually decreases with
increasing area; light falling on the outermost part of the receptive
field of an 'on-off' ganglion cell can prevent a response which
would have occurred if this part of the retina had not been
illuminated.

An inhibitory zone similar to that of the 'on-off' cells of the frog
surrounds the receptive fields of all or nearly all of the ganglion
cells of the light-adapted cat (Kuffler, 1953; Barlow, FitzHugh &
Kuffler, 1957; Wiesel, 1960). In the cat, illumination of the central
part of a cell's receptive field causes either an on-response only or
an off-response only. Those cells that give on-responses to
illumination of the central part of the receptive field give off-
responses to illumination of its surround and vice versa. If both
are illuminated together, there is little or no response. The centre
of a receptive field in the cat (i.e. excluding the opposite acting
surround) is circular, and its diameter varies between ⅛ and 2 mm,
i.e. between about 0·5° and 2° of visual angle. The size of the
dendritic trees of retinal ganglion cells in the cat (Brown & Major,
1966) agrees closely with the size of the centres of the receptive
fields. The surround is much larger, and weak excitatory effects
can be obtained with moving stimuli some tens of degrees away
from the centre of the receptive field (McIlwain, 1964, 1966).
Barlow, FitzHugh & Kuffler (1957) found that when the eye is
fully dark-adapted, the surround effects disappear, and all lights
that have any effect on the cell have the same effect. This change
in organization occurs only when dark-adaptation is nearly complete. It is not linked to the Purkinje shift, and measurements of
the relation of threshold to area with red and blue stimuli prove
that it does not depend on the change from cones to rods as the
receptors responsible for exciting the ganglion cell.

The spatial properties of ganglion cells in the spider monkey

(Hubel and Wiesel, 1960; Gouras, 1967) are very similar to those found in the cat.

*Temporal properties of ganglion cells.* Responses of ganglion cells to flickering light have been examined in the cat by Enroth (1952), who found that the flicker fusion frequency of a ganglion cell was closely correlated with the initial spike frequency that it gave in response to a single stimulus, the fusion frequency being usually between one-quarter and one-tenth of the initial spike frequency. Other properties of ganglion cells under flicker have been extensively studied, but little of interest inferred from them; references are given by Ogawa, Bishop & Levick (1966).

*Sensitivity of ganglion cells to movement and orientation.* Hartline (1940b) observed that some optic nerve fibres of the frog respond to movement of a spot of light within their receptive field. Barlow (1953b) found that this holds only for 'on-off' fibres, and that pure 'off' fibres give no response unless the movement takes the spot of light into a less sensitive part of the receptive field. It is the failure of the pure 'off' fibres to respond that is the more remarkable, since it implies that increased illumination of some receptors can exactly compensate for the effects of decreased illumination of others. There is an obvious analogy with the compensation between receptors of different kinds discovered by Donner & Rushton (1959a; see p. 86).

In the rabbit, Barlow & Hill (1963) and Barlow & Levick (1965) have found that some ganglion cells respond selectively to movement in one direction. The preferred direction is the same for all parts of the receptive field, and the same for dark spots on a bright background as for bright spots on a dark background. Ganglion cells (or optic nerve fibres) specifically sensitive to linear stimuli of certain orientation, and quite insensitive to the perpendicular orientation, have been found in the frog by Maturana, Lettvin, McCulloch & Pitts (1960) and in the rabbit by Levick (1967). In the striate cortex (see p. 117), sensitivity to one direction of movement or to one orientation has been found in all species investigated, but its occurrence in retinal ganglion cells seems to be restricted to certain species. There is no indication that any ganglion cells in the monkey have directional or orientational sensitivity, and in the cat such cells are rare: Stone & Fabian

(1966) found one among 50 cells of the area centralis, and none have yet been reported in peripheral retina. Ganglion cells of the cat may, however, have other specializations; for example Rodieck (1967) found that a small minority of them have a resting discharge that is not increased by any stimulus whatever, but is decreased by any spatial contrast appearing anywhere within a receptive field of about 2° diameter.

# 4

# THE CENTRAL PATHWAYS OF VISION

## The afferent fibres of the optic nerve

WE have seen how the retina transmits its message towards the brain in the form of trains of identical impulses in the million or so fibres of the optic nerve. Nerve fibres are efficient structures for the transmission of information from one place to another without loss. It is probably very rare, except at a junction or sudden increase in diameter, for an impulse that starts at one end of a fibre to fail to reach the other, or for a meaningless impulse to arise spontaneously in an intermediate segment. It may be that the only significant loss of information in transit is that which results from the lower velocity of propagation of an impulse that follows immediately behind another. A train of closely but unequally spaced impulses arrives at the far end of the fibre with the intervals more nearly equal, so that if the details of the spacing between impulses were used to convey significant information, much of the information would disappear. This defect of the nerve fibre as a transmission line provides one of several good reasons for believing that the nervous system does not, in general, code its information for transmission in the form of a fine pattern of short and long spaces between impulses. Probably the exact value of the interval between an impulse and its nearest neighbour is of little importance, but the time occupied by a group of half a dozen or so impulses is highly significant. For information of the latter kind, the nerve fibre may well be an almost loss-free transmission line.

The majority of the fibres of the mammalian optic nerve pass, after decussation of certain fibres in the optic chiasma, to the lateral geniculate nucleus of the thalamus. It is generally believed that in man the fibres that decussate are all those from the nasal half of each retina and none from the temporal half, the only exceptions being some of those fibres whose ganglion cells lie very near to the boundary between the two halves. Clinical observations on patients with lesions of the optic tract strongly support this belief. In the cat, however, Stone (1966) has given good evidence that, besides all or nearly all fibres from the nasal half of each retina, about a quarter of those from the temporal half decussate.

From the lateral geniculate nucleus, information is relayed mainly, perhaps wholly, to the striate cortex of the posterior pole of the cerebral hemisphere. There is, however, a substantial minority of fibres of the optic tract, estimated at 20–30% in man by Bernheimer (1899), that run to other parts of the brain. Most of these non-geniculate fibres end in the superior colliculi and pretectal region. A few, called 'accessory optic fibres', go to a wide range of other places; their course and distribution are discussed, with full bibliography, by Hayhow, Webb & Jervie (1960).

Some optic nerve fibres may send a branch to the lateral geniculate nucleus and another branch to some other part of the brain. I can find no published estimate of how many do so, except in the rat, in which Sefton (1968) showed elegantly by electrophysiological means that the great majority of retino-collicular fibres are collaterals of retino-geniculate fibres, the branching occurring at some point central to the optic foramen.

*The functions of the non-geniculate fibres of the optic tracts*

It is commonly supposed among clinical neurologists that patients in whom the occipital poles of the cerebral hemispheres, including all the striate cortex, have been destroyed are permanently and totally blind, even though the non-geniculate fibres of the optic tracts are intact and their principal further connexions undamaged. Published cases of cortical blindness are few, but support the common clinical opinion quite well. In two recent cases that were very carefully examined for traces of residual vision (Brindley, Gautier-Smith & Lewin, 1969), the only trace that could be found was the ability to distinguish sudden darkening of a light room from sudden lightening of a dark one. In one of these cases, cortical blindness, total with the above exception, has now

(December, 1969) lasted more than 3 years. The common clinical belief is thus at least nearly correct: when the cortical projection of the geniculate fibres has been lost, the non-geniculate fibres cannot easily be shown to perform, in man, any function beyond causing the pupils to constrict in response to light and perhaps allowing the subject to distinguish sudden light from sudden darkness.

It seems surprisingly wasteful that so many fibres should be reserved exclusively for so simple a function, the information for which could readily be carried in one-thousandth of the number of fibres known to be present, and it seems very probable that some additional visual function for the non-geniculate fibres of the human optic tracts remains to be discovered. In lower vertebrates, the roof of the mid-brain is the principal receiving area for fibres of the optic nerve, and very complex responses to visual stimuli can be made in the absence of the 'higher' parts of the brain. A frog with the telencephalon removed can catch flies, and a pigeon subjected to the same ablation (thalamic preparation) avoids obstacles in walking, and can be provoked to fly accurately from one perch to another (Schrader, 1889), though it does not fly spontaneously. In mammals, nothing as complex as this occurs. No response to visual stimulation can be detected in thalamic rabbits, dogs and cats except constriction of the pupil and narrowing of the palpebral fissure. But after ablations restricted to the striate cortex and its neighbourhood in cats, dogs and monkeys, though blindness may seem complete at first, a capacity to make quite complex visual discriminations returns after some weeks (Winans, 1967; Humphrey and Weiskrantz, 1967; Pasik, Pasik & Schilder, 1969; and earlier writers to whom they refer). We have to conclude either that man differs from other mammals in this respect, or that the experimental lesions in animals did not remove the whole geniculo-cortical projection area, or that the lesions causing cortical blindness in man encroached upon regions outside but near the geniculo-cortical projection area that are somehow necessary for the utilization of non-geniculo-striate visual information. Pasik & Pasik (1968) have found that the capacity for visual discrimination of monkeys with the striate and neighbouring cortex removed is abolished by lesions that interrupt the accessory optic fibres to the nucleus paralemniscalis, but probably leave the connexions of the superior colliculi intact.

*Superior colliculi.* The ipsilateral half of the contralateral retina is mapped on each superior colliculus by direct connexions, and the ipsilateral halves of both retinas by indirect connexions through the striate cortex (Wickelgren & Sterling, 1969 and earlier work to which they refer). The two maps are at least roughly concordant.

It has long been thought likely that information carried by optic tract fibres to the superior colliculi can be used in controlling the movements of the eyes. Thus Adamük (1870) showed that eye movements could be produced by stimulating the superior colliculi electrically. Apter (1946) found that after strychnine had been applied to a point on one superior colliculus in a lightly anaesthetized cat, diffuse illumination of either eye caused conjugate deviation of the eyes towards the part of the visual field that projects to the strychninized point on the colliculus; and Smith and Bridgeman (1943) showed that optokinetic nystagmus (movement of the eyes to follow a moving pattern occupying all or most of the visual field) could be abolished in guinea-pigs by collicular lesions.

The matter, however, is not clear. Optokinetic nystagmus is disturbed not only by collicular lesions but also by lesions of the visual cortex. The disturbances produced by destruction of the colliculi are very transient, disappearing completely in monkeys, within 2–10 days (Pasik, Pasik & Bender, 1966); those produced by ablation of all the striate cortex last, according to Pasik, Pasik & Krieger (1959), at least eight months in monkeys, and the second patient of Brindley, Gautier-Smith & Lewin (1969) had no trace of optokinetic nystagmus two and a half years after infarction of the posterior poles of both hemispheres, though there was good clinical evidence that the mid-brain was undamaged.

Disorders of vision produced by lesions restricted to the superior colliculi have been reported by Denny-Brown (1962) and Anderson & Symmes (1969) for monkeys and by Blake (1959) and Sprague & Meikle (1965) for cats, though denied by Pasik, Pasik & Bender (1966) for monkeys and Myers (1964) for cats. The disagreement probably depends on the extent of the lesion, and I am inclined to accept the view of Myers and of Pasik, Pasik and Bender that in cats and monkeys lesions that leave the tegmentum of the mid-brain intact have little or no visual effect. But this hardly diminishes the interest of the positive finding: a lesion of one superior colliculus, accompanied by some damage to the underlying tegmentum but without any injury to the geniculo-striate system, causes a cat or monkey to fail to react to objects presented in the contralateral half of its visual field. This clearly establishes some involvement of either the superior colliculus or the underlying tegmentum (and very probably both) in responses to just that visual information (i.e. from the contralateral half-field) that is transmitted by non-geniculate fibres of the optic tract to the colliculus. The failure of the cats and monkeys to react cannot be due to a disorder of eye movements, even if there is such a disorder (which is doubtful), for when the stimuli are presented in the intact half of the visual field the animals respond by trunk and limb movements before they move their eyes.

The disorders of vision produced by lesions of the visual cortex and of the superior colliculi are different in kind, as has been most clearly shown in hamsters by Schneider (1969). Ablation of the visual cortex on both sides caused great impairment of pattern discrimination, but hardly affected the hamsters' ability to make visually guided movements towards a seen object. After collicular lesions, the hamsters failed to move towards interesting objects (sunflower seeds) presented within the visual field, but discriminated patterns normally.

The responses of single cells in the superior colliculus to variously patterned stimulation of the eyes has been examined in the rabbit (Horn & Hill, 1966) and the cat (Straschill & Taghavy, 1967). In both species, directional selectivity is prominent, as it is in optic nerve fibres of the rabbit but not those of the cat. The most striking feature of collicular responses is their fatiguability: a series of identical stimuli gives a diminishing series of responses, and ultimately often no response. The original sensitivity returns in less than a minute.

*The pretectal region.* The one known function of non-geniculate fibres in the human optic nerve is to control the size of the pupils. The only fibres necessary (Bechterew, 1883; Ranson & Magoun, 1933) are those that runto the pretectal region, where they synapse with cells whose axons run equally to the ipsilateral and contralateral oculomotor nuclei.

*Accessory optic tracts.* The responses of single fibres of the accessory optic tracts to light falling on the retina have been studied in the rabbit by Walley (1967). The receptive fields tend to be larger than those of retino-geniculate fibres, but are not otherwise noteworthy.

Massion & Meulders (1961) and others have found that electrical activity in response to light falling on the eyes can be recorded from a wide variety of sites in the brainstem. Much of this activity is still present in cats from which the geniculo-cortical projection areas and neighbouring cortex have been removed a few weeks before, so it almost certainly depends on non-geniculate fibres. There is no evidence to decide whether these fibres are accessory, or collicular, or (least likely) pretectal. Some of the activity may be concerned in non-specific arousal of the brain, and in controlling, through the hypothalamus, the release of gonadotrophins (Clark, McKeown & Zuckerman, 1939; Critchlow, 1963), but it seems unlikely that these are its only functions.

## The lateral geniculate nucleus

The fibres of the optic nerve that supply visual information to the cerebral cortex do not pass directly to it, but synapse with cells in the lateral geniculate nucleus of the thalamus (also called the dorsal lateral geniculate nucleus, or the dorsal nucleus of the lateral geniculate body; the ventral nucleus is a small and obscure structure). Axons of these cells run, as the geniculo-calcarine tract, to the occipital cortex. The geniculo-calcarine tract is probably the only output from the lateral geniculate nucleus, but the optic tract is not its only input. There is excellent anatomical evidence for cortico-geniculate fibres, mainly from regions of cortex near the striate, and less abundantly from the striate cortex itself (Szentagothai, Hamori & Tömböl, 1966; Garey, Jones & Powell, 1968). The effects of stimulation of the mesencephalic reticular formation on transmission across geniculate synapses (Suzuki & Taira, 1961; Hotta & Kameda, 1964) make it probable that there are other inputs besides.

It is interesting to speculate on what may be the function of the lateral geniculate body, and to consider whether an animal gains any advantage by having synapses and cell-bodies intervening between its optic and geniculo-calcarine tracts, or whether, on the contrary, the visual pathway would work equally well if the fibres of the optic nerve continued without interruption to the cortex.

The principal function of the nerve cells of the retina is probably to compress the information contained in the activity of a very large number of receptors into a much smaller number of channels, the fibres of the optic nerve. In doing this they certainly discard a great deal of information. Presumably most of this loss is necessitated by the compression into fewer channels, and an animal that evolved a means of achieving the same compression with less loss of information would benefit from its discovery. Some discarding may, however, be advantageous, if biologically important information is to be efficiently sifted from unimportant. It is doubtless easier for the brain to deal with little information than with much, and a filter that prevents it from being told things that it never needs to know may usefully simplify its task.

By analogy, two possible functions suggest themselves for the lateral geniculate body. It may compress information into fewer channels, or discard unimportant information, or both of these. A third possibility is that, without necessarily either compressing or discarding, it translates the information into a different code; a code more suitable for the operations that the cerebral cortex has subsequently to perform on it. A specific suggestion of this kind would be that instead of transmitting separately the information supplied by the two eyes, it transmits by one set of fibres the average of the two messages and by another set the difference between them. Since in the normal binocular use of the eyes the two messages are nearly the same, the information contained in the difference between them could be carried by a very small number of fibres, and an elegant economy achieved. This specific suggestion is certainly false, since human subjects presented with different images on the two retinas can obtain from each image almost as much information as if the other had not been simultaneously presented; if the lateral geniculate nucleus transmitted the average of the two messages in full, but the difference only in the incomplete manner allowed by a very small number of fibres, this would not be possible.

Although there are objections to regarding the lateral geniculate nucleus as a re-coding station making use of comparisons between the messages received from both eyes, the hypothesis that it performs some kind of translation into a code more convenient to the cerebral cortex must be considered seriously if any non-trivial function for it is to be sought. The simplest such translation would be into similar trains of impulses in fibres of a different kind. If Feldberg and Vogt (1948) were right in suggesting that the chemical transmitter by which a neuron affects others must differ from all the chemical transmitters by which it is itself affected, such an apparently trivial translation may be important. A more complex kind of useful translation might involve the use of stored information, the code being changed according to the kind of message that has passed through during the preceding seconds or minutes. In some part of the visual pathway stored information must be used in such a way, to account for the after-effect of seen motion, though we shall argue in Chapter 6 that this probably occurs in the cortex rather than in the lateral geniculate nucleus.

We have thus to consider what is known anatomically and physiologically about the lateral geniculate nucleus in relation to the following hypotheses:

1. In transmitting visual information from the optic nerves to the cortex it may usefully (*a*) discard that which is unimportant, or (*b*) compress into fewer channels, or (*c*) change the code, perhaps using stored information to determine the manner of its re-encoding.

2. The non-visual inputs to the lateral geniculate nucleus could be used to regulate what information is discarded and how the rest is re-encoded. As Tello (1904) suggested, such regulation might be useful in connexion with changes in the state of attention.

## Anatomy

Our principal interest is in man and in the apes and catarrhine monkeys whose visual anatomy and physiology very closely resemble his; but because nearly all the physiological information that we have concerns it, we shall need also to consider the cat.

*Laminar structure.* The lateral geniculate nucleus of primates and of the cat consists of a number of thin layers containing cell-bodies and nuclei of nerve cells, alternating with layers which contain only nerve fibres and a few neuroglia cells.

In the cat there are three cellular layers, of which the most superficial is usually called A, and second $A_1$, and the third B. If one optic nerve is cut, and the cat killed several months later and the brain examined, many of the nerve cells are found to be reduced in size and impaired in staining properties in layers A and B of the lateral geniculate body of the opposite side and in layer $A_1$ of the same side (Minkowski, 1920). This is trans-synaptic atrophy, a phenomenon not certainly known to occur in any other part of the central nervous system. Besides its three layers, the cat's lateral geniculate nucleus has two small separate groups of nerve cells, the medial and central interlaminar nuclei.

In primates there are six cellular layers, usually numbered from 1 to 6, beginning with the most superficial (i.e. ventral). The cells of layers 1 and 2 are substantially larger and less numerous than those of the other layers. After section of one optic nerve, atrophy is found in layers 1, 4 and 6 of the lateral geniculate nucleus of the opposite side and in layers 2, 3 and 5 of that of the same side (Minkowski, 1920; see Fig. 4.1). This atrophy is not complete; Goldby (1957) found that in a man from whom one eye had been removed 40 years before he died, about half the nerve cells of the atrophic layers were missing. The surviving cells of the atrophic layers were, on the average, much smaller than cells of the corresponding layers in normal subjects. It is uncertain whether this is because the large cells disappear, leaving only small ones, or because the survivors shrink; the latter is perhaps more likely.

Removal of the occipital cortex of one cerebral hemisphere of a primate causes all the nerve cells of the lateral geniculate nucleus on the same side to degenerate (Polyak, 1933; Polyak & Hayashi, 1936). This degeneration has usually been interpreted as purely retrograde, and as implying that all cells of the primate lateral geniculate nucleus send axons to the occipital cortex. However, there is abundant evidence (Tello, 1904; Cajal, 1904; O'Leary, 1940; Szentagothai, Hamori & Tömböl, 1966) that in the cat many cells of the lateral geniculate nucleus (Golgi Type 2 cells) have short axons that are restricted to the nucleus itself. The electron-microscopical structure in primates (Colonnier & Guillery, 1964) is so similar to that in the cat that it can hardly be doubted that cells with short axons occur in primates also; and if so, the degeneration that occurs after removal of the occipital cortex cannot be purely retrograde.

*Point-to-point projection of the retina.* The retina is mapped in a regular manner on the lateral geniculate nucleus. The earliest

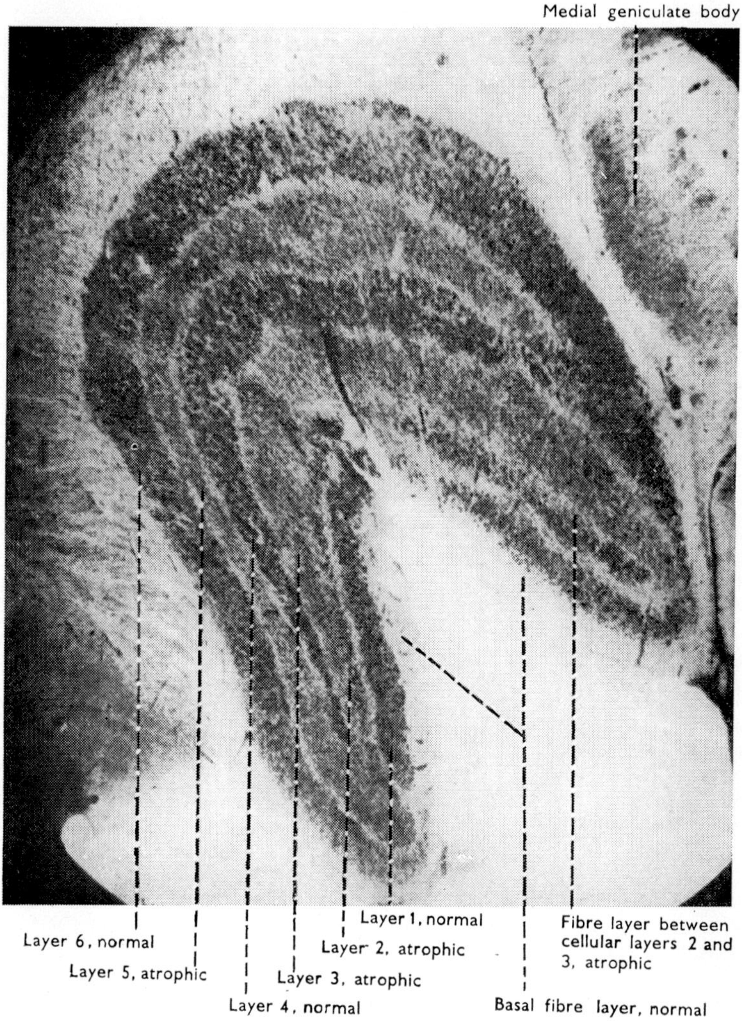

Medial geniculate body

Layer 6, normal
Layer 5, atrophic
Layer 4, normal
Layer 3, atrophic
Layer 2, atrophic
Layer 1, normal
Fibre layer between
cellular layers 2 and
3, atrophic
Basal fibre layer, normal

FIG. 4.1. Section through the left lateral geniculate body of a
rhesus monkey, 8 months after enucleation of the left eye (from
Minkowski, 1920).

[*facing p. 98*]

evidence suggesting this came from Henschen's description (1898) of blindness in the lower left quadrant of both visual fields produced in man by a lesion confined to the upper half of the right lateral geniculate nucleus. The details of the projection were worked out by Brouwer & Zeeman (1926), who made small incisions in the retinas of monkeys, and subsequently examined Marchi preparations of the lateral geniculate bodies for degenerating fibres. Clark & Penman (1934) refined the technique by looking not for degenerating fibres but for trans-synaptic atrophy of geniculate cells. They found that, if they made very small retinal incisions, the trans-synaptic atrophy could be restricted to a few hundred cells, forming compact groups in neighbouring parts of layers 1, 4 and 6 of the contralateral lateral geniculate nucleus and layers 2, 3 and 5 of the ipsilateral.

Similar compact groups of degenerated cells, but involving neighbouring parts of all six layers of one lateral geniculate nucleus were found by Polyak (1933) in monkeys in which small lesions had been made in the striate cortex. It is clear from this anatomical evidence, from the scotomata produced in man by small lesions of the visual pathway, and from electrophysiological evidence, that fibres that arise close to each other in the retina usually end close to each other in the lateral geniculate nucleus, and fibres that arise close to each other in the lateral geniculate nucleus usually end close to each other in the cerebral cortex. This organization is often called the point-to-point projection of the retina on the cortex, but it ought not to be assumed from the name that there is no overlap in the projection. The geniculo-cortical fibres entering a given 'point' (say of 100 µm diameter) in the cortex receive information from a region of retina which certainly also influences geniculo-cortical fibres entering neighbouring cortical points and probably to a small extent (Brindley & Lewin, 1968) fibres that enter quite distant points on the cortex.

There is some controversy as to the concordance of the maps of the two retinas on the lateral geniculate nucleus and on the cortex. For the cortex, the matter will be considered later: it suffices for the present to say that the maps are certainly at least nearly concordant in man, monkeys and cats; the controversy concerns only fine details. For the geniculate nucleus, the disagreement is much deeper. The maps of the contralateral retina on layers 1, 4 and 6 and of the ipsilateral retina on layers 2, 3 and 5 seem from the work of Brouwer & Zeeman (1926) and Clark & Penman (1934) to be concordant in monkeys, and the observations of Polyak (1933) on the effects of small cortical lesions indirectly confirm at least a rough concordance; yet Glees, Hallerman & Naeve (1964), in the same species and by techniques that should be at least as good, found that the upper half of each retina projects to the medial half of the contralateral lateral geniculate nucleus and to the lateral half of the ipsilateral, i.e. that there is a very large departure from concordance. The three early papers agree so well with each other that I suspect an error in the recent one, but can find no likely source of such an error.

The mapping of the retina on the cat's lateral geniculate nucleus has been very thoroughly examined by electrical recording (Seneviratne & Whitteridge, 1962; Bishop, Kozak, Levick & Vakkur, 1962) and by staining for preterminal degeneration (Stone & Hansen, 1966). On the main laminas there are concordant maps of the contralateral half of the visual field (i.e. of the ipsilateral half of the ipsilateral retina on layer $A_1$, and of the ipsilateral half of the contralateral retina on layers A and B). The maps preserve spatial continuity both from retina to geniculate and from geniculate to retina. There is a non-concordant map of part of the contralateral half-field on the medial interlaminar nucleus.

*Internal synaptic arrangements.* O'Leary (1940), after a careful study of Golgi preparations in which many fibres could be followed over a large fraction of their intrageniculate course, concluded that in the cat the terminal branches of any one fibre of the optic tract are restricted to a single lamina of the lateral geniculate nucleus. The conclusion is not necessarily contradicted by the physiological evidence that some geniculate cells in the cat can be influenced through either optic nerve, for one eye might influence such a cell through processes of geniculate cells connecting one lamina with another (though O'Leary saw no such processes). For primates, there is little anatomical evidence on which to decide whether each optic tract fibre supplies only a single lamina.

Electron microscopy has provided important information about the details of synaptic structure. In the cat (Szentagothai, Hamori & Tömböl, 1966; Peters & Palay, 1966), all optic tract axons end in special structures called synaptic glomeruli. Each glomerulus has outputs of only one kind, i.e. dendrites of geniculo-cortical

relay cells. Inputs of three kinds have been demonstrated: optic
tract axons, cortico-geniculate axons, and the axons of Golgi
Type 2 cells; the possibility of other inputs is not excluded.
Within each glomerulus, all the three known inputs synapse with
branches of output dendrites. The glomeruli also contain many
axo-axonal synapses; of the six kinds of these that are conceivable
with three kinds of axon, all occur except the two in which the
optic tract axon is post-synaptic.

There are many synapses in the lateral geniculate nucleus
besides those of the synaptic glomeruli. These probably include
synapses of cortico-geniculate cells on to geniculo-cortical relay
dendrites, and of collaterals of the geniculo-cortical relay axons
on to dendrites of Golgi Type 2 axons.

There is little anatomical evidence to indicate how many
different optic tract fibres synapse with each geniculo-cortical
relay cell, or how many geniculo-cortical relay cells receive
synapses from each optic tract axon; it can be argued from the
appearance of Golgi preparations that both numbers are unlikely
to be less than three or more than a hundred.

In the first edition of this monograph I reviewed published
counts of cells and fibres, and concluded that in the rhesus monkey
and cat there are more cells in the lateral geniculate nucleus than
fibres in the optic nerve, and in man and the rat slightly fewer. But
such estimates are very uncertain; all that can be said confidently
is that the numbers are of the same order.

*Physiology*

*The response to electrical stimulation of the optic nerve.* If one
optic nerve of an anaesthetized cat is stimulated with single
shocks, an electrode inserted into the lateral geniculate nucleus
records, after each shock, an electrical response that consists of
an initial positive-negative diphasic wave, representing the action
potential of the fibres of the optic tract, and a later negative wave
representing that of the geniculate cells.

A single shock to the optic nerve does not always cause only a
single spike in each post-synaptic neuron that is affected; repetitive firing is common. The lighter the anaesthesia and the more
'normal' the preparation, the more likely is the radiation spike
to be followed by a train of up to ten further small spikes (Bishop,
Jeremy & McLeod, 1953). The pattern of repetitive firing is not

the same in all geniculate cells. Burke & Sefton (1966a) were able to distinguish, in the rat, two classes of cells. The commoner, called P cells, give a single spike as their sole immediate response to an optic tract volley; the less common, called I cells, give an immediate burst of about ten spikes within the first 30 msec. Both kinds of cells often respond also with bursts delayed by 100–300 msec or more. Burke and Sefton present good indirect evidence that the P cells are geniculo-cortical relay cells, and the I cells are interneurons, i.e. Golgi Type 2 cells.

After a single optic tract volley, transmission through the lateral geniculate nucleus is severely depressed for over 100 msec (Marshall, 1949). This depression has been analysed by Burke & Sefton (1966b, c). It affects both P and I cells, and can be provoked not only by an optic tract volley but also by a shock to the ipsilateral visual cortex. Burke and Sefton argue that the I cells are activated by recurrent collaterals of the axons of P cells, that the axons of I cells make inhibitory synapses with P cells, and that all the depression by optic tract volleys and a substantial part of that produced by shocks to the cortex depends on postsynaptic inhibition of P cells by impulses coming from I cells. Their argument is well supported by their findings that intracellular records from P cells show a hyperpolarization (inhibitory postsynaptic potential) that agrees in time-course with the depression of conduction, and that there is no rise in the excitability of optic tract terminals within the geniculate nucleus during the depression of conduction that follows an optic tract volley. There is such a rise (though it is small) during the depression of conduction that follows a shock to the visual cortex. Burke and Sefton therefore accept that in the rat, as more conspicuously in the cat (see below), there is a presynaptic element in the second kind of depression.

It was discovered independently by Angel, Magni & Strata (1965), Iwama, Sakakura & Kasamatsu (1965) and Suzuki & Kato (1965) that stimulation of the visual cortex or mesencephalic reticular formation in the cat causes, with similar time-courses, depression of transmission through the lateral geniculate nucleus, increased excitability of optic tract fibres within the nucleus, and a slow electrical wave recordable from the optic nerve. These are properties commonly accepted (following Wall, 1958) as characteristic of presynaptic inhibition. Together they suffice to convince me that presynaptic inhibition occurs in the lateral geniculate nucleus, despite the difficulty of finding an anatomical basis for it. The combination of depressed transmission through the nucleus and enhanced excitability (presumably due to depolarization) of optic tract terminals occurs during certain eye movements. This was first observed by Sakakura & Iwama (1965) in the eye movements of the 'desynchronized' or 'rapid eye movement' phase of sleep, and later by Kawamura & Marchiafava (1968) during tracking eye movements. It may possibly be related to the suppression of vision that occurs during human saccadic eye movements (see p. 186), but Cohen, Feldman & Diamond (1969) looked unsuccessfully for it during saccadic eye movements in monkeys.

*The response to light falling on the eye.* If the eyes of an anaesthetized cat are stimulated with a brief flash of light, an electrical change, slow in comparison with the spikes described in the last section, can be recorded with an electrode in the lateral geniculate nucleus (Hunter & Ingvar, 1955). It consists of a negative wave of about 100 µV lasting about 20 msec, followed by a slower positive wave of smaller amplitude. It probably arises mainly from geniculate cells, but in part from optic tract terminals. Little has been learned from it about the functioning of the nucleus; the examination of single geniculate cells has been more profitable.

The activity of single geniculate cells has been recorded extracellularly in cats (Hubel & Wiesel, 1961; Kozak, Rodieck & Bishop, 1965) and in monkeys (De Valois, 1965 and earlier; Wiesel & Hubel, 1966). The responses of these cells are strikingly similar to those of retinal ganglion cells; in cats and monkeys the receptive fields are circular and usually have concentrically arranged mutually antagonistic regions, one responding when the light is switched on and the other when it is switched off. There is little indication of recoding, or of the discarding of any information present in the discharges of optic nerve fibres. Interaction between the two eyes is very slight; in the cat, most cells of layers A and B respond only to illumination of the contralateral eye, and most cells of layer $A_1$ only to illumination of the ipsilateral. There are, however, a very few cells that can be activated from either eye, and a larger number (perhaps one in five) which, though they respond only to illumination of one eye, do so in a manner that varies according to whether the other is illuminated. In monkeys, binocular interaction is even less, and may be absent.

The cat's lateral geniculate cells, like its retinal ganglion cells, discharge spikes spontaneously in darkness. Hubel (1960) found that in unanaesthetized cats the pattern of this spontaneous discharge differs between the waking state, where spikes come singly, and sleep, where they come in groups separated by intervals of 200 msec or more. The spontaneous discharge of some lateral geniculate cells persists after destruction of both retinas (Bishop, Burke & Davis, 1962).

In the rabbit, where most retinal ganglion cells are directionally selective, it is not surprising that directional selectivity is found also in most lateral geniculate cells. But Levick, Oyster & Takahashi (1969), found that the kind of directional selectivity is different. A retinal ganglion cell is strongly excited by movement in the preferred direction, and unaffected or weakly excited by movement in the opposite direction; a lateral geniculate cell is strongly excited by movement in the preferred direction but *inhibited* by movement in the opposite direction.

In cats anaesthetized with chloralose (Hotta & Kameda, 1964) and in unanaesthetized rabbits (Kuman & Skrebitskiĭ, 1968) some geniculate cells have been found to respond to non-visual stimuli, in particular sounds and shocks in the radial nerve. More interestingly, the receptive fields of geniculate cells for photic stimuli can be altered by stimulation of the skin (Godfraind & Meulders, 1969) or mesencephalic reticular formation (Meulders & Godfraind, 1969).

Among mammals, only the higher primates are known to have a good capacity to discriminate colours, and their colour vision system, as assessed by training methods, is very similar to the human, so that the coding of colour in their visual pathways is of special interest. It has been examined with spatially diffuse illumination by de Valois (1965 and earlier) and with controlled patterns of illumination by Wiesel & Hubel (1966).

In the small-cell layers (3 to 6 of Minkowski) most cells have receptive fields with a centre-surround organization, the centre and surround being circular and concentric and the spectral sensitivities different for the centre and the surround. Smaller classes of cells have opponent-colour responses but no centre-surround organization, or centre-surround organization but no opponent-colour responses. The main class is divided by Wiesel and Hubel into five subclasses. In order of abundance, they are red-on-centre, red-off-centre, green-on-centre, green-off-centre, and blue-on-centre. If the centre gives on-responses, the surround gives off-responses; if the centre is sensitive to red or blue, the surround is sensitive to green, and if the centre is sensitive to green, the surround is sensitive to red. It is not established that the subclasses are homogeneous in their spectral sensitivities, and it cannot be assumed, on the common and well-supported view (see pp. 222–223) that there are three kinds of cones, red-sensitive, green-sensitive and blue-sensitive, that, for example, a red-on-centre geniculate cell is connected only to red-sensitive cones in the centre of its receptive field and only to green sensitive cones in the outer part of its receptive field, though this is likely to be at least approximately the case for at least some such geniculate cells.

In the large-cell layers (1 and 2 of Minkowski), most cells have receptive fields with centre-surround organization but the same spectral sensitivity for centre and surround. This spectral sensitivity varies from cell to cell, but never, among the rather small number examined, corresponds to what would be expected of a cell connected only to a single type of cone.

When the eyes are dark-adapted, some but not all geniculate cells (probably less than half) improve greatly in sensitivity, lose their centre-surround organization, and change their spectral sensitivity to that of rhodopsin. Other geniculate cells behave as if they had no potential connexions from rods, i.e. they improve only slightly in sensitivity and retain the spatial and spectral properties that they had in the light-adapted state.

## Fibres carrying impulses towards the retina

### Anatomical evidence

A considerable body of anatomical evidence has been collected which suggests that in the optic nerves of many vertebrates there may be some fibres whose cell-bodies lie not in the retina but somewhere in the brain. Cajal (1889), in Golgi preparations of the retinas of birds, described fibres of the optic nerve which penetrated the inner plexiform layer and ended by dividing into numerous small branches amongst the amacrine cells. No connexion with any cell-body in the retina could be detected. Similar structures were later described by Polyak (1941) in the chimpanzee.

In the same year in which Cajal first inferred the existence of optic nerve fibres with centrally placed cell-bodies from observations on the retina, von Monakow (1889) drew the same conclusion from the observation that many of the cells of the superficial grey matter of the anterior colliculi of the rabbit degenerate after enucleation of one eye. It was assumed by von Monakow that these must be the cells of origin of fibres of the optic nerve, the alternative possibility that the degeneration was trans-synaptic being at that date unfamiliar and lacking in plausibility.

Recent evidence leaves no doubt about the existence of centrifugal fibres to the retina in the pigeon. Cowan & Powell (1963) showed that they arise from cells of the isthmo-optic nucleus, a small nucleus in the dorso-caudal midbrain, and run as a compact and well-defined bundle, the isthmo-optic tract, to join the optic tract close to the chiasma, in which they all decussate. Holden (1968) has recorded from single cells of the isthmo-optic nucleus, and has shown convincingly that many of them can be activated antidromically from the optic nerve. There is a rough retinotopic projection (at least quadrant-to-quadrant) of the retina through the tectum on to the isthmo-optic nucleus, and a roughly concordant projection of the isthmo-optic nucleus on to the retina (McGill, 1964). The centrifugal fibres enter the inner nuclear layer of the retina and synapse with amacrine cells and perhaps also with displaced ganglion cells (Dowling & Cowan, 1966). The number of centrifugal fibres is estimated by Cowan and Powell at 1% of the fibres of the optic nerve.

In mammals, the matter is less certain. In favour of centrifugal fibres there are the old observations of Cajal and Polyak. More recently, there are the finding of Wolter & Liss (1956) that the shrunken optic nerves of people from whom an eye had been removed many years earlier still contained structures that seemed to be nerve fibres, and the experiments of Brooke, Downer & Powell (1965), who cut one optic tract or optic nerve intracranially, and after only 6 days in cats and 7 days in monkeys found degenerating nerve fibres in electron micrographs of the retina. None of these constitutes compelling evidence in favour of centrifugal fibres, and there is some evidence against. In the cat, a few optic nerve fibres remain normal in appearance 12 weeks or more after enucleation of the corresponding eye (van Crevel, 1958). But equally many remain if, when the eye is enucleated, the optic nerve is also cut through where it enters the chiasma (Brindley & Hamasaki, 1966); there is thus no reason for supposing that the long-surviving fibres have their cell bodies in the brain. Further (Brindley and Hamasaki), after a cat's optic nerve has been cut through intracranially no degeneration is visible in Nauta preparations of its intraorbital part up to the ninth day, though after enucleation of the eye many fibres show signs of degeneration as early as the fourth day. The technique used by Brindley and Hamasaki might fail to detect degenerating unmyelinated fibres if there were any, but according to Wendell-Smith (cited by Donovan, 1967) the cat's optic nerve contains no unmyelinated fibres.

## Physiological evidence

In the fish *Ameiurus nebulosus*, whose rods, cones and pigment epithelial cells, like those of many other poikilothermic verte-brates, show photomechanical responses, Arey (1916) showed that the responses are lost in an eye whose optic nerve has been cut through. If only half of the nerve is cut through, the loss of photomechanical responses is limited to the half of the retina supplied by the cut half of the nerve. Here it is fairly clear that efferent fibres in the optic nerve must be capable of causing the movement of receptors and pigment cells. The appropriate stimu-lus is presumably light falling on either eye. The arrangement seems curiously uneconomical. The frog (Fujita, 1911) has a less surprising mechanism, the photomechanical responses of each eye depending only on light that falls on the same eye.

For mammals, physiological observations of four kinds have been interpreted as indicating that there are centrifugal fibres to the retina.

1. *Electrical stimulation of the optic nerve or tract or of some part of the brain has some effect on the retina.* For example, Motokawa & Ebe (1954) found that the sensitivity of retinal ganglion cells to light was affected by electrical stimulation of the optic nerve, and Granit (1955) that it was affected by electrical stimulation of the tegmental reticular formation. Dodt (1956) and Ogden & Brown (1964) found that after a single electrical stimulus to the optic nerve or tract the antidromic volley that could be recorded from the retina was followed by other electrical activity whose properties suggested that it was postsynaptic. Such phenomena

may well depend on centrifugal fibres if these exist; but they are of little use as evidence for their existence, since the possibility that they might depend on antidromic impulses in sensory fibres is not excluded. For effects of stimulation in the brain, the sympathetic system is another possible pathway (Mascetti, Marzi & Berlucchi, 1969*a*, *b*; see below).

2. *A natural stimulus to a sense organ other than the corresponding eye causes spikes in the optic nerve.* This was found by Spinelli, Pribram & Weingarten (1965) for clicks and for vibratory stimuli to the skin. The observations are highly suggestive of centrifugal fibres, but the possibility that the impulses observed were antidromic in sensory fibres, and of no functional significance, remains open. Of all natural stimuli, clicks and vibrations are those most likely to provoke the sort of activity, simultaneous in very many neurons, that might be expected to spread inappropriately, either by ephapsis or as a by-product of the action of axo-axonal synapses.

Cavaggioni (1968) showed that arousal by auditory stimuli causes an increase, lasting a few seconds, in the frequency of impulses in the optic nerve. But it was shown by Mascetti, Marzi & Berlucchi (1969*a*) that if the optic chiasma is cut through (so that each optic tract receives only from the ipsilateral eye) unilateral cervical sympathectomy abolishes the effect of auditory stimuli on the ipsilateral optic tract, but not on the contralateral. It seems that the influence on the retina of arousal by auditory stimuli is conveyed by sympathetic fibres, not by centrifugal fibres of the optic nerve. Mascetti, Marzi & Berlucchi (1969*b*) showed that electrical stimulation of the cervical sympathetic trunk (cut below to avoid reflex effects) augments the activity of optic nerve fibres in a corresponding manner.

3. *Section of the optic nerve affects the sensitivity of the corresponding eye to light.* Jacobson & Gestring (1958) reported that in lightly anaesthetized cats and monkeys section of the intracranial part of one optic nerve caused an immediate increase in the amplitude of the electroretinogram of the eye whose nerve had been cut. The finding in monkeys is not of much value as evidence, nor would that in cats have been if the optic nerve had been cut in the orbit, because in either case the blood supply to the eye might have been affected by cutting the nerve; but intracranial section of the optic nerve in cats escapes this criticism completely, since the ophthalmic artery of the cat is a branch of the external carotid.

Thus the existence of tonic centrifugal influences on the cat's retina through fibres of the optic nerve could hardly be doubted if Jacobson and Gestring's observation were correct. However, Brindley & Hamasaki (1962) were unable to confirm it, either in lightly anaesthetized or in unanaesthetized cats.

4. *A natural stimulus to a sense organ other than the corresponding eye affects the sensitivity of the retina.* In cats, clicks have been found to affect the amplitude of electroretinogram for a given flash (Spinelli, Pribram & Weingarten, 1965), and a flash of light to the left eye to affect the optic nerve discharge produced by a flash to the right eye (Haft & Harman, 1967). Both observations are rather thinly documented, and that of Haft and Harman was made on one cat only. However, if they can be confirmed and extended, they seem to provide the best chance of firmly establishing centrifugal influences, and of beginning to discover their significance in the life of the animal that has them.

*Function of centrifugal fibres to the human retina, if they exist*

If I take a more sceptical attitude towards the question of centrifugal fibres to the mammalian retina than the reader may think appropriate, it is in part because of the difficulty of finding any convincing suggestion of how the centrifugal control of retinal activity postulated could be of sufficient use to an animal to justify the occupation of so much valuable space in the overcrowded optic nerve, and because of the complete lack in the great body of information that we have about human vision, of any phenomenon that requires centrifugal control for its explanation. It is the general experience of those who have investigated human visual thresholds that visual sensitivity in man is not greatly altered when the subject's attention is diverted by means of stimuli applied to other sense organs. The careful experiments of Mertens (1956) in which the frequency of detecting a very dim flash of light when the subject knew where it was to appear was compared with the frequency of detecting it when it was presented at random in any one of four places, showed a very small and only just significant difference. Illumination of one eye has either no effect (Crawford, 1940; Mitchell & Liaudansky, 1955) or only a very small one (Helms & Prehn, 1958) upon sensitivity to a flash delivered to the other, when the influence of pupillary constriction is excluded. If the sensitivity of the retina could be affected by the withdrawal of attention or other activities of the brain to an extent likely to be significant for survival, some at least of these or related experiments should give clear and substantial positive results, instead of the minimal and ambiguous results that they do give. Even if clear positive results were obtained, they would not demand centrifugal control of the retina for their explanation, since the discarding of information could be carried out in the central visual pathways.

## The cortical receiving area for vision

The axons of the cells of the lateral geniculate body form in man a well-defined tract which, after curving round the inferior horn of the lateral ventricle, enters the posterior pole of the cerebral hemisphere and ends in the occipital cortex, mainly, but probably not entirely, in the part structurally defined as striate. Localized lesions of this geniculo-cortical tract, or of the optic tract or striate cortex, cause compact defects of the visual field which are closely similar, though not identical, for the two eyes; evidently fibres conveying information from corresponding points of the two retinas remain close to each other, and to fibres conveying information from neighbouring points.

The manner in which the retinas are projected on the cerebral cortex has been determined by studying the field defects caused by war wounds of the occipital poles of the hemispheres (Holmes, 1918; Teuber, Battersby & Bender, 1960) and by examining the sensations caused by electrical stimulation of the cortex (Foerster, 1929; Brindley & Lewin, 1968). From war wounds, Holmes derived the map shown in Fig. 4.2. Electrical stimulation confirms Holmes's map, and shows (Fig. 4.3) that there is a second map that is superimposed on the first and is roughly its reflexion in the horizontal meridian.

In experimental animals, the projection of the retina on the cortex has been investigated by stimulating small regions of the retina by means of a spot of light projected on to a screen in front of the animal, and recording electrical responses with an electrode on the surface of the cortex. In monkeys, Talbot & Marshall (1941) found a single projection on the occipital cortex, agreeing roughly with what had been inferred for man from the effects of lesions, but more extensive on the lateral surface and less so on the medial surface, facing the falx. In the cat, a second visual area was found by Talbot (1942), and a third visual area by Hubel & Wiesel (1965), who showed that the first, second and third areas corresponded to the histologically defined areas 17, 18 and 19 respectively (see Otsuka & Hassler, 1962). The three projections in the cat have been mapped in detail by Bilge, Bingle, Seneviratne & Whitteridge (1967). The visual input to the second and third visual areas of the cat comes in part from the first visual area, as Hubel & Wiesel (1965) showed histologically, but in part by direct fibres from the lateral geniculate nucleus, as Berkley, Wolf & Glickstein (1967) showed by examining the effects of lesions on electrical responses.

Besides the three contiguous projections of the retina that correspond to the histological areas 17, 18 and 19, there are in the cat other regions of the cortex that give large electrical responses to stimulation of the eye or optic nerve. The best known of these is on the lateral bank of the suprasylvian gyrus (Clare & Bishop, 1954). This and several others were mapped by Vastola (1961) and Bignall, Imbert & Buser (1966), who showed that they remained responsive after removal of most of the deep structures of the fore-brain medial to the lateral geniculate nuclei, and of the first and second visual areas; it is thus likely that they receive fibres directly from the lateral geniculate nuclei. The visually responsive area on the lateral bank of the supra-sylvian gyrus, however, also receives visual input indirectly, through fibres that come from area 17 of the cortex (Clare & Bishop, 1954; Hubel & Wiesel, 1965, 1969b).

*Electrical activity of the visual cortex*

*The response to electrical stimulation of the visual pathway.* If an approximately synchronous volley of impulses is sent through afferent fibres, or through commissural fibres of the corpus callosum, into a typical sensory area of the cerebral cortex, for example the somatic or the auditory receiving area, a surface-positive wave of about 100 μV lasting about 6 msec, followed by a slower surface-negative wave of variable and usually smaller amplitude, can be recorded between an electrode placed on the pial surface and a remote indifferent electrode. This is called the 'primary' cortical sensory response, and for a given preparation and arrangement of electrodes it is very constant. It is not much

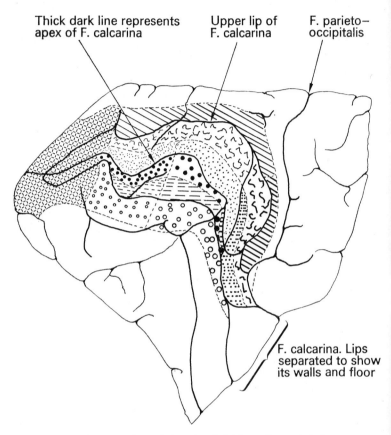

Thick dark line represents apex of F. calcarina

Upper lip of F. calcarina

F. parieto-occipitalis

F. calcarina. Lips separated to show its walls and floor

FIG. 4.2. The mapping of the visual field on the occipital cortex, as inferred from the field defects caused by war wounds (from Holmes, 1918).

affected by depth of anaesthesia or by repetitive stimulation at rates of up to about 5 per sec. It may be followed by a much more variable 'secondary' response. Records of the response of the visual cortex to electrical stimulation of its afferent pathway were first obtained as early as 1933. In the cat, the response has a very characteristic form (Marshall, Talbot & Ades, 1943; see Fig. 4.4). Its initial surface-positive part consists of four distinct waves, the second smaller than the first, third and fourth. Many attempts have been made (Schoolman & Evarts, 1959; Widén & Ajmone

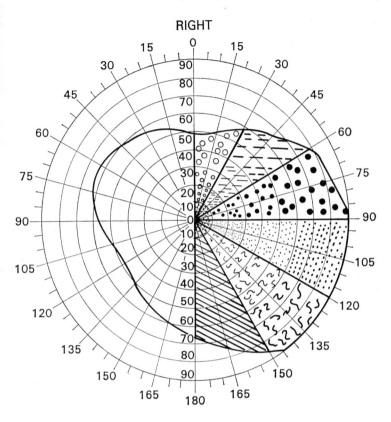

FIG. 4.2 *contd*. The parts of the visual field that project to regions of cortex shown in the drawing on the opposite page.

Marsan, 1960; and others to whom they refer) to discover which of the four waves represent spikes in incoming fibres and which are postsynaptic. But no one criterion for their presynaptic or postsynaptic nature is decisive, and different pieces of evidence conflict. It can fairly safely be concluded that the first surface-positive wave is presynaptic and the fourth postsynaptic, but the nature of the second and third remains uncertain.

Dumont and Dell (1958) studied the influences of stimulation of the ears and the reticular formation upon responses of the visual

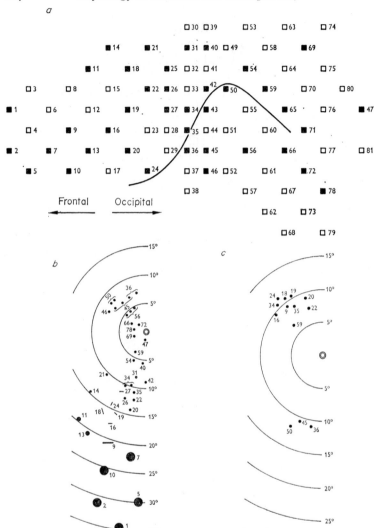

FIG. 4.3. The two maps of the visual field on the occipital cortex found by electrical stimulation. (a) The array of electrodes placed against the medial and posterior surfaces of the right occipital lobe; the curved line shows the conjectured position of the calcarine fissure. (b) The positions of the low-threshold phosphenes given by 35 of the electrodes. (c) The positions in the visual field of the high-threshold phosphenes given by 12 of the electrodes. The map of high-threshold phosphenes resembles that of low-threshold phosphenes inverted about the horizontal meridian (from Brindley & Lewin, 1968).

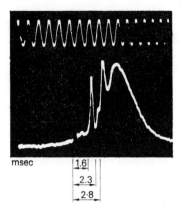

FIG. 4.4. The early response of the cat's striate cortex to a single volley of impulses in the optic tract. Downward deflexion represents positivity of the surface of the cortex in relation to a remote 'indifferent' electrode (from Marshall, Talbot & Ades, 1943).

cortex to single shocks applied to the optic chiasma. They found very large effects upon the fourth positive wave and the late negative wave, and smaller but still clearly significant effects upon the first three positive waves. The effect on the first positive wave shows that the lateral geniculate body can be influenced by pathways other than the optic nerve. The effects on the later parts of the response probably depend on reticulo-cortical pathways. Closely analogous phenomena are known in the auditory and somatic sensory pathways.

The response shown in Fig. 4.4 is followed, about 20 and 150 msec. later by further electrical responses (Bonnet & Briot, 1969), which coincide with periods of facilitation of the early response to a second shock. The response at 150 msec is abolished by precollicular transection of the brainstem.

*The response to light falling on the eye.* The response recorded by an electrode on the pial surface of the striate cortex when light is suddenly allowed to fall upon the eye consists of a positive-negative diphasic complex, usually followed by a second slower surface-positive wave (Fischer, 1932; see Fig. 4.5). When the light is suddenly extinguished, an 'off-effect' almost indistinguishable from the 'on-effect' is obtained. The initial diphasic

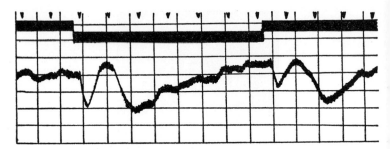

FIG. 4.5. Electrical response of the striate cortex of a curarized rabbit to light falling on the contralateral eye. Downward deflexion represents positivity of the surface in relation to a remote 'indifferent' electrode. Time-marks 0·2 sec. The stimulus-marker, immediately below the time-marks, falls during illumination. The amplitude of the responses is about 500 μV (from Fischer, 1932).

wave is several times slower than the otherwise roughly similar response obtained with electrical stimulation of the visual pathway. Records of the same kind can be obtained from the occipital scalp of conscious human subjects, if the responses to a large number of similar stimuli are averaged, to improve the discrimination from irrelevant cortical activity (Calvet, Cathala, Hirsch & Scherrer, 1956). Such averaging has become a fashionable research technique. In the principal European and American journals for the years 1964 to 1969 there are at least sixty papers devoted to averaged photically evoked activity recorded from the scalp, but I can find among them only one discovery that clearly increases knowledge of how the visual pathway works. This is that in some situations where perception of a stimulus to one eye is suppressed in binocular rivalry (Cobb, Ettlinger & Morton, 1967; MacKay, 1968), the electrical response from the occipital scalp is also suppressed; the suppression is thus occurring at a fairly early stage in the processing of visual information.

Responses to illumination, recorded with electrodes very large in comparison with nerve cells, have provided a useful means of defining the limits of the cortical receiving area for vision and of investigating the projection of the retina upon it; but they do not take us much further. To understand what the visual cortex does, we must learn how single cells of it respond. Appropriate records were first made by Jung, von Baumgarten & Baumgartner (1952), and first extended to patterned stimulation of the retina by Hubel (1958). Only a minority of cortical cells respond at all to diffuse illumination; most require some kind of pattern, and for some the range of effective patterns is very restricted.

In the cat and rhesus monkey (Hubel & Wiesel, 1959, 1962, 1968) the receptive fields of units of the striate cortex (area 17) are of four kinds, called circular-concentric, simple, complex and hypercomplex. Units with circular-concentric receptive fields behave very similarly to geniculate cells. They are found, in monkeys, only in layer 4B of the cortex, and may not be cells, but geniculo-calcarine fibres. The other three kinds of units are certainly cortical cells. Those with simple receptive fields differ from retinal ganglion cells or geniculate cells in that the excitatory and inhibitory (or on-excitatory and off-excitatory) parts of the receptive fields are not circular but elongated, and are usually not concentric but lie side by side. Otherwise, they resemble retinal ganglion cells; in particular, their response to combinations or sequences of stimuli can be predicted in a fairly simple way from their responses to the stimuli given singly. Linear stimuli that are parallel to the long axis of the elongated receptive field tend to be more effective than similar stimuli in other orientations, but only to the extent that they may occupy a large fraction of the on-excitatory field and none of the off-excitatory or vice versa. For moving stimuli, certain directions may be more effective than others, but only to the extent that these directions are associated with movement from the off-excitatory to the on-excitatory part of the field or vice versa.

Cells with complex receptive fields share, in extreme form, the property of those with simple receptive fields that appropriately orientated linear stimuli are especially effective. The special features of a complex cell are that its response to an appropriately orientated bright line or dark line or boundary between light and darkness is the same over a fairly large receptive field (of the order of 3° of perimetric angle), and that circular stimuli and inappropriately orientated lines or boundaries are ineffective, or nearly so, wherever they lie on the retina. Some complex cells are more sensitive to dark lines, some to bright lines, and some to boundaries.

Hypercomplex cells were first discovered in areas 18 and 19 in the cat, where, especially in area 19, they are very abundant

E

(Hubel & Wiesel, 1965). Later, they were found to occur also in area 17, both in the cat and in the rhesus monkey. The hypercomplex cells of area 17, and the majority of those of areas 18 and 19 ('lower-order hypercomplex'), resemble complex cells in that they respond only to an appropriately oriented line or boundary. They differ in that they respond only if the line or boundary terminates within the receptive field, or bends so that it has the appropriate orientation only over part of the receptive field. Some cells require a termination or bend at one end of the receptive field, and are unaffected by whether there is a second termination; others require a termination at both ends, or respond weakly to an appropriately oriented line that terminates or bends only at one end of the receptive field, but strongly to one that does so at both ends.

Hubel & Wiesel (1963, 1968) have examined the grouping of cells in relation to their function. The mapping of the retina on the cortex, which was already known to preserve continuity in most but not all places on a coarse scale, is found by Hubel and Wiesel also to preserve continuity on a fine scale, but with a fairly large random scatter among adjacent cells. Superimposed on this essentially continuous map they find two discontinuous mappings.

The first of these is of receptive field orientation. Cells that lie near each other in the visual cortex tend to have their receptive fields similarly oriented. Along a track perpendicular to the surface of the cortex, all cells, whether simple, complex or hypercomplex, are found to have the same receptive field orientation. Along a track parallel to the surface of the cortex, the receptive field orientation changes, in some places continuously and in others discontinuously. The assembly of neighbouring cells that share a common receptive field orientation is for historical reasons called a 'column', though it is rarely columnar in shape (Hubel & Wiesel, 1968, 1969a). The surfaces that bound it are very well-defined where the receptive field orientation changes discontinuously, less well-defined where it changes continuously.

The second discontinuous mapping is of ocular dominance. Among cortical cells of all receptive field types, some respond to stimulation of one eye only, some to both eyes equally, and some to both eyes, but unequally. Cells that lie near each other tend to have similar ocular dominance. Hubel and Wiesel have obtained evidence, partly from electrical recording and partly by anatomical means, that the surfaces that bound assemblies of cells with similar ocular dominance are everywhere perpendicular to the surface of the cortex. The assemblies are generally sheet-shaped, but for the same historical reasons as apply to the field-orientation assemblies, the name 'column' is applied to them.

Opponent-colour properties, i.e. excitation by light of some  wavelengths and inhibition by light of other wavelengths, are found in a small minority of retinal ganglion cells in monkeys, but in a majority of cells in the lateral geniculate nucleus. This trend does not continue. Among simple cortical cells those with opponent-colour properties are a minority (6 of 25 cells examined by Hubel & Wiesel, 1968) and among complex and hypercomplex cells a smaller minority (9 of 177 cells). It seems that most cortical cells respond to spatially appropriate stimuli independently of their colour, though cells that now seem indifferent to colour may perhaps prove not to be so when subtler tests are applied to them.

It is evident, and has not been neglected by Hubel & Wiesel (1962, 1968), that simple ways of connecting nerve cells can be devised theoretically that yield the simple type of cortical response-pattern by interaction of several cells with circular concentric receptive fields, the complex type of response-pattern by inter-action of several simple cortical cells, and the hypercomplex type of response-pattern by interaction of several complex cortical cells. Only one synaptic relay between each stage of complexity and the next is obligatory, though some plausible schemes use two. Denney, Baumgartner & Adorjani (1968) have made some progress towards testing these notions by first classifying cortical cells by methods similar to those of Hubel and Wiesel, and then measuring the latencies of their responses to a volley in the geniculo-calcarine tract. Units with circular-concentric receptive fields respond with short latencies that are compatible with their being geniculo-calcarine fibres, but do not exclude the possibility that some may be separated by one synapse from such fibres. Units with simple receptive fields respond with latencies longer by about one milli-second on the average. The latencies of units with complex recep-tive fields are still longer, and vary greatly from unit to unit. The measurements of latency are thus compatible with the suggestions made by Hubel and Wiesel about hierarchical organization.

Barlow, Blakemore & Pettigrew (1967) found that binocularly

driven cortical cells in the cat have just the properties that one would expect to be useful to an animal that keeps the axis of its two eyes parallel or converging on a viewed object and needs to assess distances by binocular clues. For every such cell there is a relation between positions on the two retinas that may be called 'corresponding'; two 'corresponding' images excite the cell better than either monocular image alone, but two non-corresponding images excite it *less well* than either monocular image alone. Different cortical cells differ one from another by as much as 6·6° in the horizontal relations that they accept as 'corresponding', but only by 2·2° in the vertical relations. The difference is appropriate, since horizontal disparity arises inevitably from the variation of distance of things seen, but there need be no vertical disparity if the eyes are correctly aligned. One might have expected the scatter of vertical 'correspondences' to be even smaller.

Horn & Hill (1969) have found that some cortical cells in cats alter their preferred axis of retinal image when the cat is tilted, the direction and magnitude of the alteration being such as to compensate for the tilt, so that the cells are responsive to a constant inclination of seen objects to the direction of gravity. This compensation, though functionally appropriate, seems to demand a complexity of organization much greater than one would have expected by analogy with the other tasks performed by primary visual cortex. But I can find no ground for doubting the observations, and they fit with the finding of Grüsser, Grüsser-Cornehls & Saur (1959) that the great majority of cells of the visual cortex are accessible to labyrinthine influences.

All the observations described above on the responses of cells of the striate cortex to patterned visual stimuli were made in anaesthetized animals; but Burns, Heron & Pritchard (1962) have made very similar observations on the same cells in the unanaesthetized isolated forebrain. It seems that such responses are not much disturbed by the rather light anaesthesia used by other experimenters. Anaesthesia may, however, have more effect on the responses of the visual cortex to non-visual stimuli. Such responses have been recorded, in unanaesthetized preparations, to electrical polarization of the labyrinth (Grüsser, Grüsser-Cornehls & Saur, 1959), to clicks (Jung, Kornhuber & da Fonseca, 1963) and to electric shocks to the skin of the flank, both contralateral and ipsilateral (Horn, 1965).

*Spontaneous electrical activity of the visual cortex.* The electrical oscillations at about 10 per sec ($\alpha$ rhythm) that can be recorded from the occipital scalp of conscious human subjects when the eyes are closed (Berger, 1929; Adrian & Matthews, 1934) are generated by the posterior part of the cerebral hemispheres. They can be recorded with electrodes placed directly on the surface of the exposed cortex, not only from the cortical receiving area for vision, but from a large region extending up to, but not including, the post-central gyrus (Penfield & Jasper, 1954; see Fig. 4.6). As recorded from the scalp, Adrian and Matthews found that the oscillations were usually present when the eyes were in darkness or when the whole visual fields were uniformly illuminated. They were abolished at once by any pattern or detail appearing in the central part of the visual field, but patterns in the peripheral part of the field did not necessarily prevent them. It was possible to abolish the oscillations or diminish their amplitude by a variety of non-visual activities, for example solving problems in mental arithmetic, provided that these occupied the whole of the subject's attention. To block the oscillations by means of visual stimuli, on the other hand, it was not necessary that the subject should pay any attention to the stimuli.

By presenting to the whole visual fields of both eyes a spatially uniform stimulus flickering at rates between 8 and 25 per second, Adrian and Matthews were able to provoke electrical oscillations at the frequency of the flicker. These had the same distribution over the scalp as the spontaneous oscillations occurring in darkness or steady uniform light, so it is probable that they were generated by the same part of the brain. When the frequency of flicker was very close to that of the spontaneous oscillations they were especially regular and stable; they were then undisturbed even by substantial non-uniformities of the central part of the field.

*Vascular and metabolic changes during visual activity of the cortex*

When an epileptic discharge occurs in a part of the human cerebral cortex exposed during an intracranial operation, the pulsation of the pial blood vessels in the region involved in the epileptic discharge can be seen to diminish or cease. Thermo-electric measurements show that the blood-flow through the cortex increases at the same time as the visible pulsation decreases. Both the increase in flow and the decrease in pulsation are restricted to

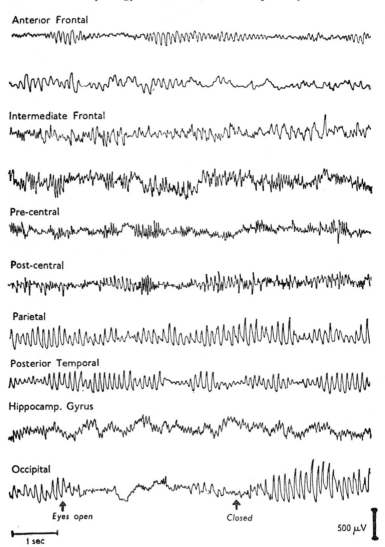

FIG. 4.6. Spontaneous electrical activity recorded directly from the exposed cerebral cortex of human subjects. Prominent rhythmic activity at about 10 cycles per second is found in the parietal and posterior temporal regions, as well as the occipital (from Penfield & Jasper, 1954).

the region of cortex involved in the epileptic discharge, and therefore cannot be secondary to any changes that may occur in the
systemic blood pressure (Penfield & Jasper, 1954). The above
observation provides the most direct evidence available that the
blood-flow to the human cerebral cortex can alter locally in
response to local activity, but in this case it is activity of a very
abnormal kind. Elegant indirect evidence had been provided much
earlier for normal activity, in a simple study by Fulton (1928)
of a patient with an angioma affecting the left occipital cortex.
The tumour produced an incomplete right homonymous hemianopia. When the patient was at rest with his eyes closed, Fulton
could hear, through a stethoscope applied to the left side of the
occiput, a faint murmur. When the patient used his eyes for
reading, or attempted to see things on his partially blind side, the
murmur became much louder. Other kinds of mental effort,
including straining to hear faint sounds, had no effect, and merely
shining a light into the eyes was similarly ineffective. The most
successful way of causing the murmur to become louder was to
ask the patient to attempt to read in a light too dim for the letters
to be easily distinguishable. The increase in murmur had a latency
of about 20 seconds, and continued for about a minute after
reading had ceased. It was largely independent of changes in blood
pressure, and there can be little doubt that it indicated a local vasodilatation provoked by activity in the underlying cortex.

The cerebral blood vessels are very sensitive to the concentration of carbon dioxide and oxygen in the blood (Wolff & Lennox,
1930), and their reaction to these chemical agents is probably a
direct one. It is thus likely that there is a close relation between
Fulton's observation and that of Meyer, Fang & Denny-Brown
(1954) that the oxygen tension in the occipital cortex of rhesus
monkeys, measured polarographically, is sharply diminished for a
few seconds by a short train of flashes of light falling on the eyes.
Activity of the cortex presumably increases its consumption of
oxygen and its output of carbon dioxide, and this in turn may
cause local vasodilatation.

*Visual sensations produced by electrical stimulation of the human
cerebral cortex*

Many observations have been made, chiefly by Foerster in
Breslau and Penfield in Montreal, on the motor and sensory effects

of electrical stimulation of the exposed cerebral cortex of conscious human subjects during neurosurgical operations (Foerster, 1929; Foerster & Penfield, 1930; Penfield & Rasmussen, 1952). The observations are not always easy to interpret, because the pattern of activity provoked in a region of cortex by electrical stimulation is probably very different from any normal pattern of activity, and it is uncertain which structures in the region stimulated are affected by the applied electric current.

For points near the calcarine fissure, the sensations produced ('phosphenes') were described by Foerster (1929) and Krause & Schum (1931) as white, point-like, and stationary in the visual field. Brindley & Lewin (1968), in a patient with an implanted array of cortical stimulating electrodes, each connected to a radio receiver beneath the scalp, found that they were always white and always flickered, even when the frequency of stimulation was above 1000 pulses per second. They were very stable in their relative positions in the visual field and probably also in their absolute positions, which were always in the contralateral half-field (see Fig. 4.3). They were usually point-like, but occasionally consisted of short lines, pairs or small groups of points, or clouds. By stimulating several cortical points simultaneously, it was possible to build up patterns in a regular and predictable way.

The findings of Penfield & Rasmussen (1952, pp. 135–144) on many patients, each one of whom could necessarily be investigated only for a short time, seem to differ. Their patients reported coloured phosphenes more often than white ones, and often described them as moving. The *movement* may not be a discrepancy, for the patient of Brindley and Lewin reported phosphenes as moving whenever she moved her eyes; they moved in the direction of the eye movement, and through roughly the same angle as that through which the eyes moved. The *colours* mentioned by Penfield and Rasmussen's patients have no counterpart in earlier or later reports.

For points remote from the calcarine fissure and probably in areas 18 or 19, the phosphenes reported by most patients of Penfield and Rasmussen were very similar to those that they reported for area 17: simple in shape and sometimes point-like, often moving, and often coloured. Three cases have been reported (Cases 4 and 9 of Foerster & Penfield, 1930, Case R. W. of Penfield & Jasper, 1954) in which stimulation of points in area 19 caused organized visual hallucinations of people, animals, and other complex shapes. In each case, the hallucinations provoked were closely similar to those that the patient experienced as part of his epileptic attack. There is no reason to expect that electrical stimulation of area 19 in subjects without epileptic foci would cause organized hallucinations.

### The degree of specialization of the visual receiving area for analysing visual information

There is much old evidence to suggest that the visual receiving area and the cortex immediately adjacent to it are especially apt to be thrown into activity by visual stimuli, perhaps more by stimuli that require analysis than by simple or meaningless stimuli that do not. The more recent experiments with microelectrodes, principally those of Hubel and Wiesel, show that the suggestion was correct, as far as rather simple kinds of analysis are concerned. There is little to indicate whether more complex kinds of analysis—the next in order of complexity might be the recognition of similarity between two retinal images related by magnification—are performed in the visual receiving area and its near neighbourhood. Experiments in which local intracortical connexions were cut probably show that these analyses *can* be performed elsewhere. Sperry, Miner & Myers (1955) found that in cats an area that almost certainly included the whole of areas 17, 18 and 19 could be thoroughly crisscrossed with incisions perpendicular to the pial surface and extending through the whole depth of the cortex, without impairing the performance of such fine visual discriminations as that between an equilateral and a nearly-equilateral triangle. Three possible interpretations need to be thought of. The first is that pathways wholly outside the area sliced by Sperry, Miner and Myers might suffice for the discrimination; if so, cats in which the same area had been removed instead of sliced should have performed as well. Though Sperry, Miner and Myers did not directly do this control, I think it is probably safe to dismiss the first interpretation; the pattern discriminations that other experimenters have found cats to perform successfully after removal of areas 17, 18 and 19 (see Winans, 1967) were much cruder. The second possible interpretation of the experiments of Sperry, Miner and Myers is that their slicing leaves the large-scale analytical functions of areas 17, 18 and 19 mainly intact because the necessary interconnexions can be made by fibres that make short loops into the white matter and back again; but this has little plausibility, as such fibres are scarce. The third and most likely interpretation is that the large-scale analysis on which the discriminations presumably depend can be performed outside areas 17, 18 and 19. Such extra-striate analytical mechanisms must almost certainly be concerned in the accurate description and discrimination of patterns of electrical stimulation of the visual cortex performed by the patient of Brindley & Lewin (1968); one would expect the electrical stimuli to interfere severely with the local small-scale analysis. But in the experiments of Sperry, Miner and Myers, the small-scale analytical mechanisms studied by Hubel and Wiesel would be mainly intact, since the slices into which Sperry, Miner and Myers cut the cortex were large compared with 'columns'.

*Possible non-visual functions of visual cortex.* It seems from the results of microelectrode recording that nearly all nerve cells of areas 17, 18 and 19 respond to appropriate visual stimuli. In unanaesthetized cats many of them respond also to non-visual stimuli, including vestibular, auditory and tactile. Some possible uses for such responses make them ancillary to visual function; for example vestibular inputs might be used only in correcting visual information for the effects of changes of posture on the retinal images, by means such as those suggested by the experiments of Horn & Hill (1969). But it is difficult to dismiss the responsiveness of the cells to non-visual stimuli as *wholly* ancillary to vision: one must suspect that despite the high degree of specialization of areas 17, 18 and 19 for visual function, they may be capable of quite other functions, as is suggested by the experiments of Lashley (1943) which demonstrated effects on maze-learning of lesions of the visual cortex in previously blinded rats.

### The joining of the two halves of the visual field

If the visual areas of the left hemisphere received visual information solely from the left half-retina and those of the right hemispheres solely from the right half-retina, one might expect that a decision about a pattern presented all on one half-retina might be in some way easier than a comparable decision in which the activity of both half-retinas have to be correlated. No such difference is noticeable to introspection, and measurements of the latencies for making such decisions (Brindley, Carpenter & Rushton, 1967) detect no difference.

The avoidance of such a discontinuity at the vertical meridian of the visual field doubtless depends to a small extent on overlap between the areas of retina that send fibres into the left and right optic tracts. Stone (1966) has demonstrated a narrow strip of dense overlap in the cat, and man is likely to be similar in this respect.

Over distances greater than the width of the strip that projects densely to both optic tracts, the joining of the two half-fields probably depends on connexions through the corpus callosum between areas 18 of the two sides. The existence of such connexions in area 18 and their absence in areas 17 and 19 was shown by electrical recording by Curtis (1940) and anatomically by Myers (1962). Choudbury, Whitteridge & Wilson (1965) examined in cats and baboons the electrical responses of areas 17, 18 and 19 to visual stimuli after section of the ipsilateral optic tract. They found that only area 18 and the immediately adjacent part of area 17 responded, and these only to stimuli very near to the vertical meridian of the visual field. The responses were abolished by cutting through the corpus callosum or cooling the mirror-point on the opposite hemisphere. The fibres concerned constitute a large fraction of the fibres of the posterior corpus callosum: Hubel & Wiesel (1967) found that among 34 of its fibres all but one or two could be driven by visual stimuli, and of these all but one had receptive fields that overlapped the vertical meridian or came to within a degree of it.

### Other visual functions of the corpus callosum

It is likely that, besides joining the two halves of the visual field, the corpus callosum has other visual functions, and that these are not restricted to the immediate neighbourhood of the vertical meridian. There are two groups of observations that suggest this, but neither quite suffices to prove it.

The first group concerns the interocular transfer of learned visual discriminations in cats and monkeys in which the optic chiasma has been cut through before the beginning of training with the aim of restricting the immediate visual input from each eye to the ipsilateral hemisphere. In such animals, as in wholly

intact animals, a discrimination learned with the left eye is performed equally well with the right (Myers, 1955). But if, before
training, the posterior half of the corpus callosum in a cat or the
posterior half of the corpus callosum and the anterior commissure
in a chimpanzee is cut through, transfer to the second eye fails
completely (Myers, 1959; Black & Myers, 1964). The successful
interocular transfer in animals whose corpus callosum is intact
might be supposed to be due to the formation of two memory-
traces, one in each hemisphere, at the time of training; or one
might suppose that the new visual information received in the
second hemisphere at the time of testing has to be correlated with
a memory-trace that is restricted to the first hemisphere. On either
hypothesis the transfer of information between the hemispheres
would presumably depend on fibres of the posterior part of the
corpus callosum. The first hypothesis is correct, for Butler (1968)
found that if the corpus callosum is cut through between training
and testing, transfer occurs. The question now arises whether the
fibres that transfer the information can be merely the known ones
that join mirror-points in areas 18 and are related only to the
neighbourhood of the vertical meridian, or whether they must
include fibres responsive to stimuli in other parts of the field. It
is natural to suspect the latter, but the question cannot be settled
until interocular transfer in chiasma-sectioned animals has been
tested with stimuli that are restricted to lateral parts of the field.

The second group of observations concerns a rare clinical condition elucidated by Geschwind & Fusillo (1966): right hemianopia
accompanied by alexia without agraphia. The inability to read
affects only *seen* writing; the reading by touch of words written in
embossed letters is unaffected. The patients have no difficulty in
naming objects placed in their hands, but make errors in the
naming of seen objects. These errors are roughly dysphasic in
type, but are unaccompanied by any dysphasia in purely verbal
performance. The errors of naming and the alexia are far from
being accounted for by the hemianopia, since right hemianopia in
other patients causes neither.

Geschwind and Fusillo's patient had a lesion (proved at
necropsy) affecting the splenium of the corpus callosum as well as
the left occipital pole. They argue that the condition of alexia
without agraphia occurs when the surviving right visual cortex is
disconnected from the speech centres of the left hemisphere by

damage to the corpus callosum. If one accepts this interpretation, one must wonder whether the callosal fibres concerned carry information only about the neighbourhood of the vertical meridian of the field, or also about lateral parts. Lateral parts of the field are certainly connected somehow to the speech centres, for normal right-handed people can recognize words presented tachisto-scopically in the left half of the peripheral field. It would be interesting to know whether any of this ability persists in the rare cases of alexia without agraphia.

## Visual functions of areas of cortex other than the receiving area for fibres from the lateral geniculate nucleus

### The temporal lobes

Klüver & Bucy (1939) discovered that removal of both temporal lobes from monkeys produced a characteristic complex change in behaviour. Some of its constituents, for example an increase in tameness and in sexual activity, were probably unrelated to dis-ordered visual function, but the response of the monkeys to things that they saw seemed to show that they could no longer recognize objects by visual criteria, though they could by tactile. Klüver and Bucy called this part of their syndrome 'psychic blindness'.

Attempts to analyse the syndrome of Klüver and Bucy by mak-ing less radical lesions have not clearly separated the gross kind of 'psychic blindness', conspicuous on casual observation of the animals, from the rest of the syndrome; but they have shown (Mishkin & Pribram, 1954; Mishkin, 1954) that defects in learned pattern discrimination, not conspicuous to the casual observer but easily demonstrated by formal testing, can be produced by bilateral lesions restricted to the inferior temporal cortex; lesions restricted to lateral temporal cortex or to the medial part of the temporal lobe including the hippocampus have no such effect. The defect caused by these inferior temporal lesions is specifically visual; it does not extend to the tactile discrimination of shapes or weights (Pribram & Barry, 1956) or to auditory discriminations (Weis-krantz & Mishkin, 1958). It is not accompanied by any disorder of visual acuity, visual fields, or visually guided reaching (Cowey & Weiskrantz, 1967). The function of each temporal lobe in visual pattern discrimination is mainly related to the contralateral half of the visual field, for Ettlinger (1959) found that ablation of the right inferior temporal cortex together with section of the left optic tract abolished a previously learned discrimination, but the same two lesions made both on the same (left) side did not. There are, however, some relevant connexions between the striate cortex and the contralateral temporal lobe through the posterior corpus callosum: Mishkin (1966) found that ablation of inferior temporal cortex on one side and striate cortex on the other caused a previously learned discrimination to be lost (as would be expected from Ettlinger's results), but permitted its relearning; section of the posterior corpus callosum then caused it to be lost again. As with the disturbance of transfer studied by Myers and with the syndrome of alexia without agraphia, new experiments are needed before it can be firmly decided whether the relevant callosal fibres transfer information only about the vicinity of the vertical meridian of the field.

### Areas 18 and 19

The visual effects of large lesions of areas 18 and 19 in monkeys are very slight, as Lashley (1948) showed. In man Penfield & Jasper (1954, p. 125) were unable to find any visual disorder caused by excision of areas 18 and 19 (presumably in one hemisphere). Several experimenters after Lashley have been able to detect in monkeys very slight visual effects of ablation of areas 18 and 19, but large disabilities have been found (Ades & Raab, 1949; Ettlinger, Iwai, Mishkin & Rosvold, 1968) only when such prestriate lesions have been combined with inferior temporal lesions. It seems that inferior temporal and prestriate cortex can in part substitute for one another, so that the disorder caused by lesions of the two together is more than the sum of the disorders caused by the two separately.

### The parietal lobes

Among the disorders of function that have been described in disease of the parietal lobes in man are several that involve an inability to make use of visual information. Two of these are *constructional apraxia*, an inability to copy or depict designs or objects, and *spatial disorientation*, an inability to find the way about, even in surroundings formerly very familiar. These can certainly both be produced, in severe form and with little disability of any other kind, by lesions in the posterior parietal cortex of the non-dominant hemisphere (Hécaen, Penfield, Bertrand & Malmo, 1956; Whitty & Newcombe, 1965). They commonly occur together, but either can occur without the other, and whether such separation has any consistent relation to the site of the lesion is unknown. The sufficiency of a right posterior parietal lesion for these disorders does not prove it to be necessary: constructional apraxia of very similar type often occurs in lesions restricted to the dominant hemisphere (see especially De Renzi & Faglioni, 1967). There are, however, subtle differences between the kinds of errors made in their attempts at drawing by patients with constructional apraxia from left and right lesions, as Warrington, James & Kinsbourne (1966) have shown.

Though constructional apraxia and spatial disorientation seem to be mainly visual disorders, they are probably not wholly so; and in any case the visual function that is disordered is so complex that it is difficult to see how one could begin to analyse it. A slightly less complex disorder that is more certainly specific to vision is the 'visual disorientation confined to homonymous half-fields' first described by Riddoch (1935; see Cole, Schutta & Warrington, 1962, for later cases). Riddoch's two patients both had left posterior parietal gliomas, affecting white matter as well as cortex. Their disability was that they made gross errors in estimating the direction and the distance of objects seen in the right half of the visual field, though they correctly reported the distance and direction of anything seen centrally or in the left half of the visual field. Both had intact fields as tested perimetrically during part at least of the time during which the visual disorientation was present. Their disorder is one that seems to have just enough simplicity for there to be a reasonable prospect that a careful study of similar cases may contribute something to our understanding of how the visual system works.

The effects of parietal lesions, both unilateral and bilateral, in monkeys, have been examined by Bates & Ettlinger (1960) and Ettlinger & Kalsbeck (1962). The disorders are not simple. Performance in some tests involving visual discrimination is impaired, but I suspect that analysis of the extent to which the impairment is specifically visual will be difficult.

# 5

# INTRODUCTION TO SENSORY EXPERIMENTS

THE physiology of sensory pathways differs from that of other parts of the body in that it is based not only on objective observations and measurements which can be described in physico-chemical and anatomical terms, but also on the results of sensory experiments, that is, experiments in which an essential part of the result is a subject's report of his own sensations. In the investigation of the visual pathway a much greater number and variety of sensory experiments have been performed than in any other branch of physiology, and it will be useful to make a general assessment of the place and value of the evidence from such experiments in relation to that from objective observations and measurements.

## Class A and Class B observations

The main function of science, in those of its branches that have advanced beyond the primitively exploratory stage, is the formulation and testing of hypotheses which have exact and potentially observable implications. The inescapable implications of any hypothesis can necessarily be expressed in terms which appear either in the statement of the hypothesis, or in the background of generally accepted theory assumed in conjunction with it. For physiology, the terms used in stating the theoretical background are physico-chemical and anatomical; so it would seem that no physiological hypothesis that is also stated in physical, chemical and anatomical terms can ever predict the result of a sensory experiment, in which a report of sensations is concerned. There is, however, one class of predictions that can be made if we add to our theoretical background a single hypothesis that is very difficult to doubt. The additional hypothesis required is that whenever two stimuli cause physically indistinguishable signals to be sent from the sense organs to the brain, the sensations produced by these stimuli, as reported by the subject in words, symbols or actions, must also be indistinguishable. It may perhaps be maintained that, when the terms 'physically indistinguishable' and 'sense organ' are properly defined, this statement is not a hypothesis at all, but a logical truism; however, we need not be much concerned with whether it is or is not a hypothesis, for we require only that it be acceptable, at least as long as no conflicting experimental observations are known. By accepting it, we enable ourselves to make predictions of indistinguishability, and thereby to use for testing physiological hypotheses all those sensory observations that assert merely that 'the stimuli $\alpha$ and $\beta$ under conditions $X$, $Y$ produce the same sensation' or 'the stimuli $\alpha$ and $\beta$ under conditions $X$, $Y$ produce different sensations'. In the discussion which follows, observations of these kinds will be called 'Class A observations'. They include the results of matching experiments and of determinations of thresholds, both absolute and incremental. The last two chapters of this book will be almost entirely concerned with Class A observations.

Any observation that cannot be expressed as the identity or non-identity of two sensations will be called a 'Class B observation'. Class B observations include all those in which the subject must describe the quality or intensity of his sensations, or abstract from two different sensations some aspect in which they are alike. The phenomena involved often have some objectivity, in that different observers agree, more or less, in their descriptions of them. Sometimes it is difficult to be sure that the agreement implies any more than that the observers have learned to use the same words to describe the same natural objects; but among the great number of Class B observations in the literature of visual physiology, from Purkinje's admirable and insufficiently known little book of 1823–1825 to the present day, and among those that will in the future be made, there are probably many that are capable of being used to provide evidence concerning the functioning of the visual pathway. Up to the present time they have mostly been used unsatisfactorily, and it is useful to re-examine in general terms the possible ways of using them.

## Psycho-physical linking hypotheses

If a physiological hypothesis, i.e. a hypothesis about function that is stated in physical, chemical and anatomical terms, is to imply a given result for a sensory experiment, the background of theory assumed in conjunction with it must be enlarged to include hypotheses containing psychological terms as well as physico-chemical and anatomical. These may be called *psycho-physical linking hypotheses*. The one that has already been stated on p. 133, namely that physically indistinguishable signals sent from sense organs to the brain cause indistinguishable sensations, is the most general, and at the same time the most difficult to doubt, that has yet been proposed. It seems to me that it is the only one that is at present sufficiently secure to deserve inclusion in the body of generally accepted theory. Others have been suggested, and will be briefly considered later in this chapter. They are mostly both special and uncertain, and when a deduction made from one of them in conjunction with a purely physiological hypothesis is found by experiment to be false, we are not bound to conclude that it is the physiological hypothesis that is contradicted.

## The conservative and liberal approaches to sensory experiments

The use of Class A observations as a basis for analysing the function of the eye and visual pathway is not controversial; every writer on vision admits, at least by implication, that they can legitimately be used. On the use of the kinds of observation here called Class B, there have been differences of opinion which can best be clarified and illustrated by describing two opposed points of view. The conservative opinion, in its most extreme form, is that only Class A observations are of any value, and in a discussion of visual mechanisms all Class B observations may be entirely disregarded. I cannot discover any direct published statement of this opinion, but it is approximately the attitude of the report of the Committee on Vision of the Optical Society of America (1933–1934), and it is implicit, though nowhere explicit, in the admirable papers on visual subjects of Dr W. S. Stiles. The extreme liberal opinion has been forcefully expressed by Stevens (1951). It makes no distinction between Class A and Class B, nor any similar division of sensory experiments, and holds that all sensory observations that can be made consistently, and upon which different subjects agree, should be equally admissible as means of analysing the function of sensory pathways.

For the psychologist, it is clear that the conservative view is too restrictive, and would lead, if generally adopted, to the total neglect by scientists of phenomena to which scientific methods of analysis are at least in part applicable. For the physiologist, interested in the testing of hypotheses expressed in physico-chemical and anatomical terms, the conservative approach has much in its favour: it provides a very easily applied rule for the rejection of irrelevant and inconclusive arguments. As long as the hypotheses to be tested are purely physico-chemical and anatomical in their statement, it will never reject an argument which could be absolutely compelling. Nevertheless, it will not be adopted in this book, partly because I do not intend to limit myself strictly to hypotheses expressible in physico-chemical and anatomical terms, and partly because, even for such hypotheses, Class B observations can sometimes provide valuably suggestive evidence. The next chapter will be mainly concerned with Class B observations, and in the remainder of the present chapter the means by which they may be used to test hypotheses will be discussed.

To give some such general discussion seems important because in the past very many arguments have been published, especially in colour vision, in which Class B observations have been compared with inferences which were derived from physiological hypotheses by reasoning containing unacknowledged non-rigorous steps. It would be easy but unnecessary to give a long list of examples of this fault; the list could include at least nine-tenths of those papers in which conclusions expressed in physico-chemical and anatomical terms have been drawn from Class B sensory experiments. The remaining one-tenth or less escape the criticism not because the non-rigorous steps are absent, but because the authors show that they recognize which parts of their arguments are compelling and which are not.

## How can Class B observations be used?

### Conversion to Class A observations

The most completely satisfactory way of using a Class B observation to test a physiological hypothesis is to convert it into one belonging to Class A. Given a phenomenon for the description of which psychological terms are required, we can, for example, ask ourselves under what conditions physically different stimuli will produce it to the same extent. Applications of this procedure are considered on pages 150 and 154. When such Class A experiments have been done and their implications explored, the information contained in the original Class B observation has not necessarily been exhausted, and it may still be used to test a psycho-physical linking hypothesis.

Another way of converting Class B observations into Class A is to extrapolate from them, using knowledge derived from other experiments, to conditions that come within Class A, and having thus guessed the probable result, to perform the Class A experiment and confirm that the extrapolation was valid. A good example is the 'Purkinje phenomenon' or difference between the spectral sensitivity curves of the eye in light-adapted and dark-adapted states. This was first described by Purkinje (1825) in a Class B form: a spectrum appeared brightest in the yellow region when viewed with the light-adapted eye, but in the green region when dimmed uniformly and viewed with the dark-adapted eye. From this it was natural to guess the Class A conclusion that the threshold of the light-adapted eye was lowest in the yellow (strictly yellow-green) part of the spectrum and of the dark-adapted in the green (strictly blue-green); so natural, indeed, that though the guess for the dark-adapted eye was tested and confirmed in the nineteenth century, that for the light-adapted eye remained merely a guess until the work of Kohlrausch (1931), who found it to be true.

Occasionally there is something to be gained by passing from a Class A experiment to a Class B variant of it. The different sensitivity of foveal cones to light coming from different parts of the pupil (Stiles–Crawford effect; see p. 248) is well known from matching and threshold experiments. It is, however, slow and troublesome to explore the directional sensitivity over the whole pupil by these means. Westheimer (1968b) described a simple optical trick (inspection of an out-of focus image of a point source) whereby a person can in a few seconds report, from a Class B observation, on the distribution of directional sensitivity of his foveal cones over his pupil.

### Extrapolation to hypothetical Class A observations

Sometimes, where a Class B observation strongly suggests the result that would be obtained from a closely related hypothetical

Class A experiment, it seems justifiable to draw conclusions without actually performing the Class A experiment, which thus remains hypothetical. An example of this appears in the experiments on the distribution of electrical phosphenes in the visual field described on pp. 155–157. It would doubtless be possible, at the cost of much time and trouble, to reproduce exactly by optical means the visual sensations produced by the electric currents concerned. However, we need to know only the spatial characteristics of the hypothetical photic stimulus that would do this. It is already known from other experiments that the plotting of spatial coincidences between photic stimuli is not substantially disturbed by differences of quality or intensity, so the spatial distribution can be plotted with almost any photic stimulus; it need not at all reproduce the quality of the sensation produced by the electric current. The strictly Class A experiment, if done, would almost certainly give the same result, and the closely related Class B experiment is technically very much easier to perform.

Arguments of this kind from Class B experiments seem to me to be justified provided that it is made clear where the loose step in the argument is, so that if grounds for doubting the conclusion subsequently appear we know where in the chain of reasoning to place the doubt. Nevertheless, it is more satisfactory to avoid the loose step by performing the related Class A experiment, wherever this is not too difficult technically.

*Argument through special psycho-physical linking hypotheses*

On p. 139 an argument is given in favour of the opinion that the rods and cones are the light-sensitive structures of the human eye. The argument requires that the tree-like figure seen when a moving spot of light is projected on the bulbar conjunctiva (another of Purkinje's discoveries) be due to the shadow cast by the retinal blood vessels on the receptors, and that a spatial coincidence be recognized by the subject between part of the tree-like figure and an external photic stimulus only when the image of the stimulus and the shadow of a blood vessel fall on the same receptors. This conclusion is so strongly supported by the agreement, for any one subject, of every detail of shape of the subjective figure with the corresponding part of his vascular tree as determined objectively, that it is hardly possible to doubt it. In itself it is a psycho-physical linking conclusion, not a physiological one, but in conjunction with experiments it can be used to investigate the physiological problem of the distance between the receptors and the vessels.

Arguments of this kind depend on first securely establishing a psycho-physical linking hypothesis by correlating very many features of a sensory phenomenon with corresponding features of an objectively determined one. The only satisfactory examples that I have seen or been able to devise relate to entoptic images (pp. 139–142), which are especially favourable because from a hypothesis of the origin of such an image the distribution of light on the retina can be obtained by purely physical reasoning; prediction of the sensory results of the distribution of light then usually requires only a trivial extrapolation from the results of other sensory experiments. There is no reason in principle why similar arguments should not be applied to other Class B phenomena, but for them it is difficult to see how the necessary psycho-physical linking hypothesis could in practice be securely enough established for the experiment to be a test not of it but of the associated physiological hypothesis.

### Use to test psycho-physical linking hypotheses

Psycho-physical linking hypotheses that have been hitherto proposed have mostly been either very special, with implications going only slightly beyond the experimental observations on which they were based, or general, but so indefinite as to have no inescapable implications. The introduction and testing of the former kind presents no unfamiliar problems. The latter kind I shall try to avoid. I have tried, and shall continue to try, to devise new psycho-physical linking hypotheses that are both general and definite. Like other theorists, I have not yet succeeded.

# 6

# SOME SENSORY EXPERIMENTS
# INVOLVING NEITHER MATCHING NOR
# THRESHOLD

## Entoptic images and related phenomena

VISUAL illusions that can be confidently attributed to the optical effects (absorption, refraction or reflexion) of structures within the eye are known as 'entoptic images'. Of the very many that have been described, the majority are due to the cornea, lens or vitreous. Four that are almost certainly of retinal origin will be considered here.

### The shadows of the retinal blood vessels

If a bright spot of light is projected on to the bulbar conjunctiva and caused to oscillate to and fro, the eye being otherwise in darkness, the subject sees his whole visual field filled with orange light, on which a tree-like figure of fine branching dark lines, moving with the oscillations of the spot of light on the conjunctiva, is clearly visible. The details of the figure for any one subject agree with the pattern of his retinal blood vessels as seen ophthalmoscopically, so that there can be little doubt that the figure is a result of the shadow cast by the retinal vessels on the receptors. H. Müller (1854) and König & Zumft (1894) estimated the distance of the receptors behind the vessels by comparing the apparent movement (in minutes of visual field) of the subjective figure with the distance through which the spot of light on the conjunctiva had to be moved to cause it. The result agrees well with the hypothesis that the rods and cones are the receptors. Lest this be thought too obviously true to require evidence in its support, it should be remembered that as recently as 1953 Ségal argued that the rods and cones constitute only one among several classes of light-sensitive structures in the retina, others being the synaptic terminals of the cone cells and the cells of the pigment epithelium. The evidence in favour of Ségal's hypothesis is weak; but the only direct evidence against it is that derived from the apparent movement of the shadows of the retinal vessels. Indirect and general arguments suffice to make it very probable indeed that the rods and cones are the only photo-receptors concerned in initiating the messages which pass along the vertebrate optic nerve; but it must be admitted that these indirect arguments are inconclusive when taken singly.

## Maxwell's spot and Haidinger's brushes

If a brightly illuminated piece of white paper is looked at alternately through a blue and a yellow colour-filter, a dark ring of diameter from 2 to 3° surrounding the fixation point is seen during the first few seconds of each period of looking through the blue filter (Maxwell, 1856). At the fixation point itself there is, for most subjects, a smaller dark region. The ring and dark centre together are generally known as Maxwell's spot. As the eye becomes adapted to the blue light, the spot fades and disappears, but it can be restored by re-adapting to the yellow. Red, orange, green, neutral grey or an opaque screen can be substituted for yellow. The spot is most easily seen if the light used for the contrasting adaptation has roughly the same subjective brightness as the blue. Instead of blue, any mixed light (for example, purple) containing a substantial proportion of wavelengths shorter than 500 nm can be used. In purple light the spot appears pink instead of merely dark.

There can be little doubt that Maxwell's spot is wholly or mainly due to the screening of the receptors by the yellow pigment of the macula. Its spatial distribution is appropriate; when seen in monochromatic light it appears, as it should, as a contrast purely of brightness and not of colour; and the range of wavelengths of monochromatic light in which it can be seen agrees closely with the range of wavelengths absorbed by macular pigment, whether this is estimated by spectrophotometry of excised retinas (Sachs, 1891) or extracts from them (Wald, 1945), or by objective or subjective methods in the living human eye (Brindley & Willmer, 1952; Stiles, 1953). The contrary hypothesis of Walls & Mathews (1952), that the spot depends not on macular pigment but on local variations in the relative abundance of the different kinds of receptors, not only predicts (unless complex *ad hoc* reconciling hypotheses are added) that it should appear as a contrast of colour when seen in monochromatic light, but has the inescapable implication, if there are only three kinds of receptors, that it must have the same appearance in one member of a pair of metameric lights (i.e. a pair of lights of different spectral composition which match) as in the other. This is far from the truth, as Stiles (1955) has clearly shown.

If the sky is looked at through a Nicol prism, a small yellow figure resembling an hour-glass is seen centred on the fixation point, with its axis perpendicular to the electric vector of the light transmitted by the Nicol prism (Haidinger, 1844). The isthmus of the hour-glass is difficult to see with certainty, and its boundaries are too blurred for the observer to be certain whether they are straight or curved, so that the figure can be equally well described by saying that the region of the visual field within about 1° 30′ of the fixation point is divided into four equal sectors, appearing alternately much yellower and slightly bluer than distant parts of the field. The whole array constitutes 'Haidinger's brushes'. If the Nicol prism (or piece of polaroid) is kept stationary, the brushes soon fade, but they can be restored by rotation. They occupy the same part of the visual field as Maxwell's spot and, like it, can be seen in monochromatic lights of those wavelengths that are absorbed by macular pigment. In monochromatic light they always appear as contrasts of brightness and not of colour. All these facts are consistent with the hypothesis that the brushes are due to a dichroic screening pigment which lies close in front of the macular receptors and at any one point in the retina preferentially absorbs light polarized with its electric vector perpendicular to a line joining that point to the foveal centre. Almost certainly this is the macular pigment.

Haidinger's brushes can be seen not only in plane-polarized but also in circularly polarized light. The earliest description of this that I can find is a note by Shurcliff (1955). The brushes are at about 45° to the horizon, the dark component being directed from upper right to lower left in right-handed circularly polarized light, and from upper left to lower right in left-handed. Their orientation is independent of the point of entry of the light through the pupil. They could be explained if any of the optical media of the eye were birefringent with axes vertical and horizontal. Incident circularly polarized light would then be converted into elliptically polarized light with its axis oblique to the horizon. The structure responsible is unlikely to be the lens, whose birefringence, if it had any, would be expected to depend on the point of entry of the light through the pupil; but it might well be the cornea.

## Moving bright points

On looking at a large uniformly illuminated surface, for example blue sky, snow or mist, bright spots can be seen which appear suddenly, move across the visual field for distances of a degree or two, and then disappear (Purkinje, 1823, p. 67). They tend to follow each other along the same track, and are more conspicuous during or immediately after muscular exertion. Abelsdorff & Nagel (1904) provided the following three arguments, which together make it almost certain that they are the shadows of red blood cells, or of gaps in otherwise continuous columns of red cells, moving along retinal capillaries. First, the tracks never approach closer to the fixation point than 50′, i.e. never enter the region of retina that lacks blood vessels. Secondly, the movement of the bright points is made pulsatile by light pressure on the eyeball, and abolished by firm pressure. Thirdly, they can be seen easily in deep blue or violet monochromatic light, and with some difficulty in yellow-green, but not at all in red, orange or blue-green; the wavelengths of light in which they can be seen are precisely those which are absorbed by oxyhaemoglobin. The scarcity of the moving bright points—several seconds usually elapse between successive traverses of the same track—suggests that they are gaps in columns of red cells rather than single red cells. Such gaps might be due to white cells.

## Abstraction, from two differing sensations, of an aspect in which they are alike

### The effects of duration and area on apparent brightness

When two spatially and spectrally similar flashes of light of different durations are both shorter than about 40 msec, suitable adjustment of their relative intensities allows them to be matched exactly. Experiments of this kind come within Class A, and will be considered in the next chapter. When one of the flashes is substantially longer than 40 msec, no adjustment of their relative intensities will make them indistinguishable. Nevertheless, subjects can disregard the obvious difference of duration, and adjust them until they are equal in apparent brightness; and it is found that different subjects agree fairly well in the settings they make on attempting this task. Broca and Sulzer (1902) were the first to perform this experiment. Their results are well known and often quoted, but no theorist has yet succeeded in using them to test a physiological hypothesis.

A related phenomenon, sometimes called 'brightness enhancement', is that a flickering light may appear brighter than a steady light of the same luminance (Brücke, 1864; Wasserman, 1966). The best frequency is usually between 5 and 15 c/sec; it increases with increasing luminance.

The spatial analogue of the temporal experiment of Broca and Sulzer has been performed by Willmer (1954) for small foveal fields. As with the temporal experiment, when the fields are extremely small they can be matched exactly, so that the observation belongs to Class A, though it may show no more than that the diffraction patterns on the retina caused by very small sources are independent of their angular subtense provided that the total luminous flux received remains constant. For larger fields, the observations are non-trivial and of Class B. They can no more be interpreted, in the present state of physiological theory, than can those of Broca and Sulzer.

*Heterochromatic photometry and Abney's law*

If two lights are alike in colour, then a judgment of whether or not they are equally bright is a judgment of whether they can be distinguished, and comes within Class A. If they differ in colour, it remains possible for a subject to abstract from the total quality of the two different sensations an aspect which he recognizes as brightness, and assess whether or not they are alike in this; and the assessments made by different subjects when asked to perform this task are found to agree fairly well with each other. Such assessments form the basis of heterochromatic photometry, about which there is an extensive literature on account of its technical importance. Its physiological foundation is likely to be very complex, and no satisfactory hypotheses have yet been put forward concerning it.

It would provide both a convenient aid to the technologist in extrapolating from limited data and an interesting starting-point for the speculations of the theoretical physiologist if brightness were additive, that is if the equality in brightness of two lights $A$ and $B$ necessarily implied the equality in brightness of $(A + C)$ and $(B + C)$, $C$ being any other light. Such additivity was first asserted by Grassmann (1853), as the fourth and last of his laws of colour mixture. Grassmann suggested that his fourth law was less well founded than his first three (which will be described in Chapter 8); later research has justified this opinion.

Grassmann's fourth law was re-stated, and some experimental evidence consistent with it described, by Abney & Festing in 1886. Since that year it has been generally known as Abney's law. Recent experimenters have not found it to be true except as a very rough approximation. In particular, when two complementary or nearly complementary spectral lights are mixed, the brightness of the mixture, as assessed by heterochromatic photometry, is found to be much less than the sum of the brightnesses of its constituents (Dresler, 1937; Guth, Donley & Marrocco, 1969). The discrepancy between the results of Abney and Festing and those of later workers probably represents a difference in the task that their subjects were attempting. Dresler refers to the possibility that one may be able to learn, after careful training, to assess a related quantity, namely 'luminance' as defined by the Commission Internationale de l'Éclairage, which is strictly additive, and Boynton & Kaiser (1968) find Abney's law to hold if the criterion for matching is a minimally distinct boundary between the fields. Nevertheless, there is no doubt that if subjects without special training are asked to adjust lights of different colours to equal brightness, the great majority of them make settings which depart from Abney's law in the sense that the brightness of an unsaturated binary mixture is judged to be less than the sum of those of its saturated components.

### Bezold–Brücke effect

Just as equality of brightness can be assessed with fair consistency in the presence of a difference of colour, so also can equality of hue in the presence of a difference of luminance or of saturation. If this assessment is made on spectral lights at different luminances, hues are found to be judged equal not when the wavelengths are equal, but when differences of wavelength are present which vary with relative luminance and with spectral region, but are fairly reproducible from one subject to another. This is the Bezold–Brücke effect, first discovered by Helmholtz (1867, p. 320) and von Bezold (1873), and investigated quantitatively by Exner (1902). Exner found that lights of wavelength greater than 577 nm or between 475 and 508 nm appeared to decrease in wavelength when brightened, and those between 508 and 577 nm or less than 475 nm to increase. Subsequent investigations have given substantially the same results. For very bright lights fixated for many seconds, the change of apparent hue can be great; for example red light may appear green (Cornsweet et al., 1958).

Not only luminance but also saturation can influence the judgment of hue, and Abney (1913) has described the effects of the addition of white upon the apparent hue of spectral colours.

## The after-effects of seen motion and form

### Motion: the 'waterfall phenomenon'

It was recognized by Aristotle that when one has looked for some time at moving objects, stationary objects may appear to be moving in the opposite direction. The many observations on this after-effect of seen motion published during the nineteenth century have been well reviewed by Wohlgemuth (1911), who describes it thus: 'If we look for a while at a waterfall, fixating a point on the stationary rock behind, and then gaze upon the stationary landscape, this will appear to move upwards. . . . From a moving carriage, receding objects grow smaller. If the carriage stops suddenly, these objects appear to grow larger, to swell, to come nearer. . . . In short, under certain conditions, a movement produces an after-effect which manifests itself as an apparent movement in the opposite direction.'

The after-effect of seen motion is restricted to that part of the visual field in which the motion has been seen. If the motion has been seen with one eye only, its after-effect can be observed almost as well with the other eye as with that which viewed it. This still holds if, between looking at the real motion with one eye and examining for its after-effect with the other, the first eye has been pressure-blinded (Barlow & Brindley, 1963). It is not even necessary, after viewing the moving object, to look at a real stationary one; if the eyes are closed, a rather feeble illusion of movement in the opposite direction is experienced, even though no light is reaching either retina.

Cords & von Brücke (1907) attempted to measure the after-effect by neutralizing it. As inducing stimulus, a rapidly moving endless band of alternate black and white stripes was used. After the subject had gazed for some time at a small stationary fixation point placed in front of the striped moving band, a band of ordinary graph paper moving more slowly in the same direction was substituted for the striped band by an optical device. If the speed of the graph paper was fairly high, its apparent movement was found to be in the same direction as its true movement, but slower. At very low speed, the apparent movement was opposite to the true. At an intermediate speed the band appeared, approximately, to stand still; but it seems from the authors' description that this neutralization was never quite perfect.

The after-effect of seen motion can be obtained more easily if the eyes are kept fixed, so that the image of the moving object moves over the retina, than if the eyes are allowed to follow the movement; indeed I cannot detect any trace of it if the moving field is followed during the whole period of observation except for the necessary 'saccadic' jerks to new fixation points.

The fact that the effect, when established with one eye, can be transferred to the other makes it very unlikely that it depends on a retinal process or on one occurring in the lateral geniculate nucleus. Its restriction to the part of the visual field stimulated suggests that it is a property of the striate cortex.

A class A counterpart of the after-effect of seen motion has been described by Pantle & Sekuler (1968): after adaptation to the motion, the threshold for detection of gratings of low contrast moving in the same direction is raised.

### Form: figural after-effects

A class of spatial illusions with properties closely analogous to those of the after-effect of seen motion was discovered by Gibson (1933). A simple example is shown in Fig. 6.1. If one looks for two or three minutes at the small circle in the upper part of the figure, allowing the gaze to wander round its circumference but not to stray outside it, and then turns to the parallel straight lines in the lower part of the figure, the line directly looked at and those to the right of it appear straight, but those to the left curved in the opposite direction to the large arcs in the upper part of the figure. This and similar illusions are known as figural after-effects. They

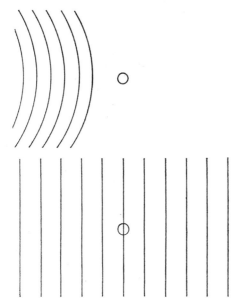

Fɪɢ. 6.1. Figural after-effect of curvature. For explanation, see text.

are confined, or almost confined, to the part of the visual field to which the inducing stimulus was delivered, but can be transferred from one eye to the other. As with the after-effect of seen motion, this suggests that the striate cortex is probably the structure whose subsequent activity is modified as a result of the inducing stimulus.

If one looks for a minute at a television screen at a distance small enough for the horizontal scanning lines to be easily resolvable, and then at a blank wall, one sees against the wall a grid of vertical lines. McCollough (1965; see also Hajos, 1969) discovered that after adaptation of the eye for 2 minutes or more to an orange-and-black vertical grating alternating temporally with a blue-and-black horizontal grating, one sees a vertical white-and-black grating as bluish and a horizontal white-and-black grating as faintly orange. Both these phenomena have obvious analogies with properties of cells of the striate cortex, and the one discovered by McCollough suggests that many cortical edge-detectors in man may be colour-specific; but it is important to keep in mind that arguments of this kind are rough and insecure.

A Class A phenomenon closely related to figural after-effects will be described on p. 169.

## Change of luminance

Anstis (1967) found that after adaptation to a saw-tooth variation of luminance (i.e. a light whose luminance alternately, over many cycles, either increased gradually and decreased suddenly or decreased gradually and increased suddenly), a light of constant luminance appeared to drift in brightness, the drift being opposite in direction to the slow phase of the inducing cycles. This after-effect is not transferred from one eye to the other; thus it may perhaps depend on properties of retinal neurons.

## The consequence of stabilizing the retinal image

If, whilst sitting still in a well-lit room, one fixes one's gaze steadily on any small conspicuous object for two or three minutes, the room soon seems to be filled with mist, and all objects in it become much less clearly visible than when the eyes are allowed to move freely from one fixation point to another. This simple experiment (Troxler, 1804) does not provide very efficient stabilization of the retinal image, for it is impossible to maintain accurate fixation voluntarily for more than a few seconds. Ditchburn & Ginsborg (1952) and Riggs, Ratliff, Cornsweet & Cornsweet (1953) devised an optical system, using a mirror mounted on a contact lens which moved with the eye, by means of which small or moderate horizontal eye movements were caused to have almost no effect on the pattern of illumination of the retina. Vertical lines were used as test objects, so that the lack of stabilization for vertical eye movements was not important. The subjects of the experiments found that during the first few seconds of examining the stabilized retinal image, the test object was as easily seen as in normal vision, but after an interval which varied with the width and contrast of the test object between about 5 and 20 seconds, it became almost or quite invisible. Often it later reappeared for a few seconds, once or several times. There has been dispute on whether the reappearances are due to accidental failures of stabilization and whether vision ultimately disappears completely; but the work of Barlow (1963) makes it clear that disappearance is complete only for fine details or features with low contrast, and that reappearances, though they can be due to failure of stabilization, are not always so.

The loss of detail and contrast when the retinal image is stabilized are probably mainly due to properties of the retina, for retinal ganglion cells are known to adapt in an appropriate way (see p. 81), and phosphenes produced by electrical stimulation of the striate cortex, though perfectly stabilized, show no loss of contrast, brightness or detail with time (Brindley & Lewin, 1968).

## Successive contrast and after-images

When, after looking fixedly at any brightly illuminated object, the eyes are closed or turned to look at a uniform surface, an image resembling in shape the original object can be seen for a period varying from a few seconds to several minutes. Such images are usually known as after-images. Some writers prefer to restrict the name 'after-images' to those seen against a zero background, those seen on a bright background being called 'successive contrast phenomena'. The two are not sharply distinguishable from each other, since dim backgrounds behave like darkness, and between dim and bright backgrounds the transitional effects are often smoothly graded. Authors who use the name 'after-images' in the wide sense often divide them into 'positive', appearing of approximately the same colour as the lights that provoked them, and 'negative', appearing of approximately the complementary colour. This distinction also is not a sharp one, since the after-image of a strong stimulus is often neither of the same colour as the stimulus nor of the complementary; it passes through a complex sequence of colours (accurately described by Padgham, 1968), which is remarkably constant from one subject to another.

A very striking phenomenon analogous to negative after-images, but not necessarily similar in its physiological basis, was discovered by Bidwell (1897) and examined quantitatively by Sperling (1960). If darkness, a coloured pattern, and a white uniform field are presented cyclically in this order so that the coloured pattern precedes the white field by about 10 msec, the colour is not seen; its complementary appears instead. Figure 6.2 shows the disc with which Bidwell demonstrated this phenomenon.

By far the greater part of the extensive literature on after-

F

FIG. 6.2. The disc with which Bidwell (1897) caused objects to appear complementary to their true colours. To demonstrate this very striking illusion, a replica of the Figure should be made and rotated clockwise in front of the objects at about 6 rotations per second.

images and related phenomena describes experiments belonging to Class B that have not yet been successfully used to test any physiological hypothesis; but it is possible to make Class A observations about them by determining the conditions necessary for physically different stimuli to give indistinguishable after-images. One group of such investigations examines whether the after-images of metameric lights can be distinguished (see p. 214). Another (Brindley, 1959a) has shown that the after-images, excluding their first fifteen seconds, produced by any fixed amount of light are exactly the same, whether the light is delivered all within 0·02 second or distributed over 2 seconds, provided that the eye is homatropinized so that the pupil does not constrict during the longer exposure. Such pairs of stimuli produce very different primary sensations and very different electroretinograms, and may be presumed from the results of experiments on animals to affect all the nerve cells of the retina differently; but some, at least, of their photochemical effects are alike. The indistinguishability of their after-images therefore suggest strongly that the lasting change responsible for the persistence of these after-images is a primary photochemical one, and does not depend on the activity of any nerve cells. This inference is supported by the fact that if two flashes of the same duration but differing by 20% in luminance are presented simultaneously to neighbouring areas of the retina, the primary sensations cannot be distinguished when the intensity of the dimmer stimulus exceeds 300 cd m$^{-2}$ sec, but the late after-images remain easily distinguishable up to much higher intensities, the discrimination failing only above about 300 000 cd m$^{-2}$ sec, an intensity at which other estimates (see pp. 12, 251) indicate that nearly all the visual pigments are bleached.

The late after-images of very bright stimuli have two other properties (Brindley, 1962b) that probably have simple photo-chemical causes. The first is that any fine detail in these after-images becomes progressively blurred during the 15–20 minutes that they can be seen. The blurring is much faster in the green negative after-images of deep red stimuli than in the pink negative after-images of stimuli of shorter wavelength. It does not occur in extrafoveal after-images, is not merely due to fading, and is independent of whether, after the stimulus, the eye is kept in darkness, in light, or in alternating darkness and light. All these properties are explained if it is supposed that the late after-images depend on product of photolysis of cone pigments which diffuse from the cones in which they are produced and diminish the sensi-tivity of other cones of the same spectral type, and that the product of bleaching the red-sensitive cones diffuses faster than that liberated from the green-sensitive. The last assumption explains also the second newly discovered property of late foveal after-images, namely that when they are produced by stimuli of long wavelengths, they are surrounded by a green halo of 0·5° to 2° diameter. Addition of light of short wavelength to the stimulus prevents the production of the green halo. The prevention may perhaps depend on uptake of the fast-moving erythrolabe product by bleached cones.

Another way of making quantitative observations about after-images is to match them with the sensation produced by light of known composition falling on an unadapted part of the same or the opposite retina. For after-images in the narrow sense, i.e. those seen against a dark background, this has been done by Barlow & Sparrock (1964). For comparison field they used a stabilized retinal image, as was appropriate, since after-images are themselves exactly stable in relation to the retina. For after-images in the wide sense, matching has been used by Schön (1874), Voeste (1898), Wright (1934) and MacAdam (1949a, 1956). Wright adapted part of the visual field of one eye to bright light of known spectral com-position, and then matched the sensation produced by subsequent exposure of this retinal region to a different and usually dimmer light against that produced by a comparison stimulus applied to a non-corresponding part of the visual field of the opposite eye. The experiments of Schön, Voeste and MacAdam were similar except that the test and comparison stimuli were applied to the same eye.

Such observations come near to Class A. They are not strictly within it only because the subject necessarily knows from the previous course of the experiment which of the sensations that he is comparing is an after-image and which is due to the comparison field. What can be inferred from the observations will be considered in Chapter 8.

## Simultaneous contrast

### Contiguous fields

A small neutral grey patch on a white background appears darker than a similar one on a black background. If it is placed on a strongly coloured background, such a grey field appears tinged with the complementary colour. These and related changes in the appearance of objects according to the brightness of their surroundings are known as simultaneous contrast phenomena. Several authors, notably Helson (1943), Fedorov *et al.* (1953), and Alpern & David (1959) have made quantitative assessments of simultaneous contrast by adjusting the physical composition of two fields until they look alike despite being surrounded by different borders. If the test fields and surrounds are presented as brief flashes, the temporal relation between them is important, the inducing field being effective only if it comes between 100 msec before the test field and 34 msec after it (Kinney, 1967). These experiments do not properly belong to Class A since the subjects necessarily know which test field is surrounded by which border. It would be advantageous to present the borders to corresponding areas of the retinas of left and right eyes, and the test stimuli to one or other eye in temporal sequence. The experiments would then come within Class A if subjects were chosen who could not, in the absence of other clues, distinguish which eye receives a flash of light on its central retina. This holds for many but not all people (Templeton & Green, 1968; Enoch, Goldmann & Sunga, 1969).

## Non-contiguous fields

A light can be caused to appear dimmer by introducing a bright 'glare' light into distant parts of the visual field. Part of this effect depends on the light reflex of the pupil, and part on simultaneous contrast of the ordinary contiguous kind induced by stray light. Whether there is also a direct effect of illumination of one area of retina on the sensation produced by a given (weaker) illumination of a distant area has never been clearly established. If there is such an effect, it is small.

## Confusion and false perception of certain distributions of luminance

Figure 6.3 shows a striking illusion discovered by T. N. Cornsweet. In Fig. 6.3(a) the central circle seems to be brighter than its surround, but is not, as can be verified by covering the boundary between them with a thin ring of paper cut to the correct size. In the similar-looking picture on the right the central circle is in fact brighter than the surround. A closely related illusion was discovered by Mach (1865; see also Ratliff, 1965): a region where the second derivative of luminance with respect to distance is negative looks brighter than its surroundings. If the first derivative is zero, such a region *is* brighter than its surroundings; but the subjective appearance is the same when the first derivative is non-zero. A region where the second derivative is positive correspondingly looks darker than its surroundings. The Mach and Cornsweet illusions depend mainly on spatial properties of the stimuli, but may have a temporal element, since the eyes are not still. They can be shown to hold for purely spatial distributions of light by viewing the Figures in a flash too brief to permit any eye movement, and for purely temporal distributions by causing a spatially uniform field of light to undergo changes in luminance corresponding to the graphs of reflectance. The illusions belong to Class B as originally described, but almost certainly have Class A counterparts: the similarity of appearance between Figs. 6.3(a) and 6.3(b) could almost certainly be made an identity of appearance without making the distributions of reflectance identical. Such 'metameric' patterns are related to the insensitivity of the eye to small uniform gradients of luminance.

## Benham's top and related phenomena

There are certain spatio-temporal patterns of illumination that provoke an impression of various colours, even though the light falling on the retina is either wholly white or wholly of a single wavelength. The necessary condition for this illusion appears to be the alternation, at a frequency a little below that which abolishes flicker, of a uniform bright field, a uniform dark field, and one containing sharp spatial contrasts of bright and dark. One means of producing such a sequence was described by Fechner as early as 1838. Perhaps the most vivid of them is 'Benham's top' (1894), which is shown in Fig. 6.4. If it is rotated clockwise at about six

FIG. 6.4. Benham's top. If the top is rotated at about six revolutions per second, the arcs appear brightly coloured.

revolutions per second, and examined in bright white light, the innermost group of three arcs appears vividly red, the next brownish, the next olive-green, and the outermost dark blue. With anticlockwise rotation the order of colours is reversed. Different observers agree well in their description of the colours, especially for the outermost and innermost arcs. Changing the speed of the top, or the intensity or spectral composition of the light used to illuminate it, alters the appearance of the colours, but even in monochromatic light the apparent colours of different arcs differ conspicuously. It was claimed by Abney (1913) and by Piéron (1929) that the apparent colours were different in different metameric lights (i.e. lights of different spectral composition that match). If correct this would be extremely important, for it would disprove the hypothesis that trichromacy depends on the existence of only three cone pigments, a hypothesis which other evidence

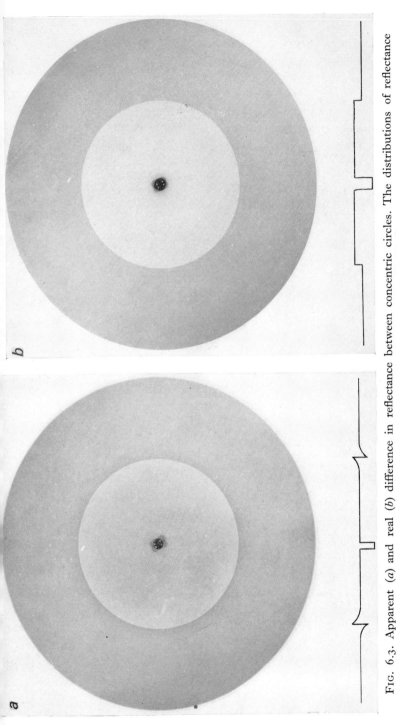

FIG. 6.3. Apparent (a) and real (b) difference in reflectance between concentric circles. The distributions of reflectance are shown below. This illusion was discovered by T. N. Cornsweet.

[facing p. 154]

very strongly supports (see Chapter 8). However, attempts to confirm the differences described by Abney and Piéron have been unsuccessful (Brindley, 1960*b*; von Campenhausen, 1969); the old observations are probably erroneous.

## Electrical phosphenes

Sensations of light ('phosphenes') can be produced not only by light falling on the retina, but also by electric currents passed through the eye. The retina is exceptionally sensitive to electricity. For example, if both eyes are temporarily blinded by pressing on them to occlude the circulation, even currents 150 times the threshold for the unblinded eye produced no phosphene (Brindley, 1955*a*); these currents are evidently still insufficient to stimulate the optic nerve. The sensitivity of the retina to electricity is not increased by dark-adaptation; on the contrary, it is decreased, though only by a factor of about 1·5.

An indication of the structures in the retina affected by electric currents near the threshold can be obtained from the distributions of the phosphenes in the visual field. The distributions for three arrangements of conjunctival electrodes are shown in Fig. 6.5. For each arrangement there is an intense sensation of light in those parts of the visual field which correspond to the regions of retina lying immediately under the electrodes. It is these regions which should be carrying the greatest density of current perpendicular to the retinal surface. Between the two electrodes there must be a zone where the retina carries no current perpendicular to its surface, but a high density of current parallel to the surface; and in the part of the visual field corresponding to the expected position of this zone there is found to be no sensation of light. This strongly suggests that the effectiveness of electric currents in stimulating the retina depends on the component of current density perpendicular to the surface of the retina, and that current parallel to the surface, whatever its relation to the direction of the optic nerve fibres, is ineffective. Such a rule would be expected if the structures affected by the current were all orientated at right angles to the surface of the retina, that is radially in the eye. Observations on the part of the phosphene close to the fixation point indicate that in the macular region of the retina the preferred direction of current for stimulating departs from the strictly radial, but remains in the

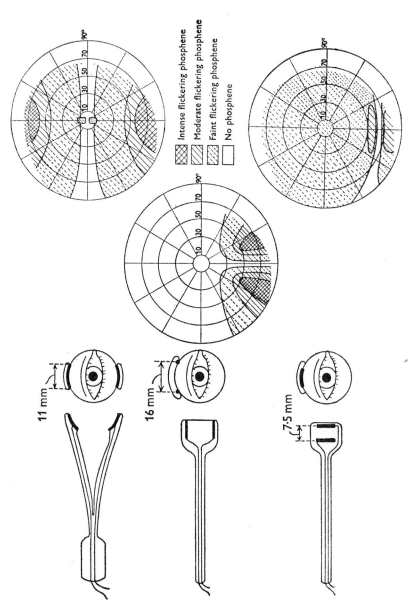

Intense flickering phosphene

Moderate flickering phosphene

Faint flickering phosphene

No phosphene

Fig. 6.5. Distribution in the visual field of sensations of light produced by electric currents whose distri-

plane defined by the visual axis of the eye and the point on the retina that is stimulated. The likely explanation is that the structures affected by the current are strictly radial throughout most of the retina, but in the macular region depart from the strictly radial in the sense of leaning outwards in all directions from the visual axis. The bipolar cells and the parts of the rod and cone cells lying inside the external limiting membrane comply with this condition. There is no firm evidence to decide which of them is responsible for the high sensitivity, but an indication may come from the observation of Trifonov & Utina (1966) that electrical stimuli to the retina resemble photic stimuli in increasing the nucleic acid content of rods and cones, as well as of nerve cells. Some other relevant considerations are discussed on p. 53.

Electrical and photic stimuli to the retina can interact in a number of ways, some of them complex and surprising. The production of beats by stimulating with flickering light and alternating electric current at slightly different frequencies will be considered in Chapter 7 in connexion with the nature of flicker fusion frequency. Some Class B observations on suppression and reinforcement of colour signals and on complex spatial effects belong properly in the present chapter.

If lights of two different colours are presented to the eye alternately in regular sequence at a frequency above about 20 c/sec, an intermediate colour is seen, roughly (at low frequencies) or exactly (at high frequencies) as if the two lights were mixed. If at the same time an alternating current of the same frequency as the light is passed through the eye, it alters the apparent colour of the light in a manner that depends on the phase relation (Brindley, 1964). The explanation is doubtless roughly that the alternating current acts somewhere along the path of conduction of the signal from the cone outer segments, so as sometimes to weaken the signal and sometimes to strengthen it, according to the direction of the current or perhaps the sign of its time-derivative. If the current is of a weakening kind during the green signal and of a strengthening kind during the red signal, it makes the sensation redder; if the reverse, it makes the sensation greener.

It has long been known (Purkinje, 1823, pp. 11–22, Smythies, 1959) that if the whole visual field is filled with light flickering at a frequency less than 40 c/sec, complex spatial patterns are seen. They are fairly constant on different occasions for the same subject and conditions, but vary greatly from one subject to another. They can be imitated closely if, instead of flickering light of a given frequency, steady light is put into the eye together with alternating current of that frequency (Wolff, Delacour, Carpenter & Brindley, 1968).

Alternating currents of frequencies greater than 40 c/sec can produce patterns that are more vivid and much more constant from one subject to another. These patterns are very beautiful. They often have bright colours and very sharp distinct outlines; they are continually changing, moving and bubbling, but always retain, for a given condition, a constant general character from which an experienced observer can diagnose fairly accurately the frequency of the current that is being passed through his eye. For the strongly, steadily and uniformly illuminated retina, currents of from 40 to 220 c/sec give patterns; for the unilluminated retina, only currents in the narrower range from 45 to 100 c/sec. Figure 6.6

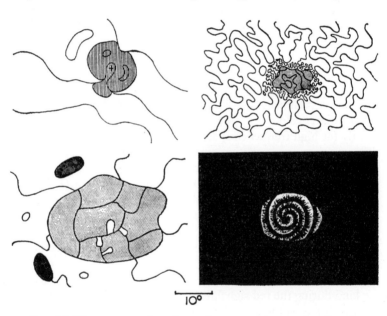

FIG. 6.6. Figures seen when alternating current is passed uniformly through the retina. (a) 75 c/sec, in uniform light of 220 cd m². (b) 110 c/sec, in uniform light of 220 cd m². (c) 130 c/sec, in uniform light of 220 cd m². (d) 60 c/sec, in darkness (from Wolff, Delacour, Carpenter & Brindley, 1968).

shows four examples. Wolff *et al.* suggest that they are likely to be generated in the retina, rather than in the brain. We have not been able to think of any clear way of testing this suggestion, or of using the patterns to provide information about how the retina works.

## The blue arcs

As Purkinje (1825, p. 74) first observed, if a small light in an otherwise dark room is viewed with the fovea, two bluish-white arcs may, under favourable conditions (for which see Moreland, 1968), be seen running from the light to the blind spot. The position of the arcs in the visual field corresponds to what would be expected if they were due to stimulation of the adjacent retina by impulses in the principal bundles of fibres running from the fovea to the optic disc (papillo-macular bundles), and other properties of the arcs leave no doubt that this is their cause. The stimulation is not by light emitted by the active nerve fibres, for no such light can be detected physically and (the stronger argument) dark-adaptation makes the arcs not easier but harder to see (Newhall, 1937). It must be the electric currents of the nerve impulses that stimulate. Whether they stimulate neighbouring optic nerve fibres or those structures (bipolar cells or the inner parts of receptors) that are stimulated by externally applied current is uncertain. Optic nerve fibres are less electrically sensitive, but on the other hand they are very much nearer. The colour of the arcs is nearly independent of that of the inducing spot of light; they are blue or violet, and fairly saturated, always much bluer than the infinite-temperature end of the Planckian locus (Newhall). Phosphenes produced by externally applied electric currents are white or only very slightly bluish; this suggests (but is insufficient to prove) that the site of stimulation is different, i.e. that the blue arcs depend on stimulation of optic nerve fibres.

## Deformation phosphenes

If, with the eyes turned far to the left, one presses with a pencil through the eyelid against the outermost accessible part of the right eye or the innermost accessible part of the left eye, one sees to the extreme left an oval dark shadow, which may have a bright border or a small bright region in the middle or both. This is a deformation phosphene.

The threshold force for producing a deformation phosphene is about four times higher in darkness than in bright light. The phosphenes produced by large forces with the eye in darkness ('positive deformation phosphenes') appear wholly bright; those produced by smaller forces with the eye illuminated ('negative deformation phosphenes') appear mainly as dark shadows.

To make a negative deformation phosphene visible it is necessary that the retina be illuminated from in front (Brindley, 1966); light that enters the retina from its choroidal surface is wholly ineffective. This proves at once that the negative deformation phosphene is due to something that prevents the normal amount of light from reaching the visual pigment. Presumably deformation interferes with the funnelling process that increases the light-capturing power of cones (greatly) and rods (slightly) for light that comes from the direction of the pupil (see p. 251). In accord with this view, the negative deformation phosphene is found to correspond to a much greater attenuation of the light in cone vision (a factor of 6) than in rod vision (only a factor of $1\frac{1}{2}$).

The positive deformation phosphene is obviously of quite different origin. It is unaffected by the adaptational state of the eye: dark-adaptation, which greatly increases the sensitivity of the retina to light and slightly decreases its sensitivity to electricity, leaves its sensitivity to deformation unchanged. The structures responsible are unknown.

# 7

# SPATIAL, TEMPORAL AND ADAPTATIONAL ASPECTS OF THE VISUAL THRESHOLD

A THRESHOLD is a stimulus such that equal or stronger ones are effective and weaker are not. Visual thresholds, defined in this way, are not sharp. There are stimuli which if repeated many times will sometimes be seen and sometimes not. We can state the frequency with which a given stimulus is detected, and use this as an estimate of the probability that it will be detected on a future occasion. In general this probability is greater for stronger stimuli and less for weaker ones. An estimate of the strength of stimulus for which it has some assigned value ($\frac{1}{2}$ is commonly used) provides an alternative and more precise definition of the threshold.

Visual thresholds may be measured against zero background (absolute thresholds) or in the presence of steady and uniform illumination upon which the test stimulus is superimposed (incremental thresholds). They are influenced by whether the eye has been exposed to the background for long enough to become accustomed or 'adapted' to it. This process of adaptation may require as long as an hour for its completion where the background is total darkness and the eye has previously been exposed to bright light, but for most other conditions it is much quicker than this. It will be convenient to restrict the first part of this chapter to observations where the eye is adapted to the background on which the stimulus is presented, reserving the important problems of what happens during the process of adaptation to later pages.

## The spectral sensitivity at threshold

When we use the extrafoveal retina adapted to darkness, we see all weak monochromatic stimuli as white or bluish-white except those of wavelengths greater than about 630 nm. If these lights of long wavelength are excluded, there is no discrimination of colours. For this and other good reasons it is generally believed that the detection of weak stimuli of colours other than red by the dark-adapted peripheral retina depends on one class of receptors only, the rods. If the number of quanta of monochromatic light which must fall on a given area of the retina in a given time in order to be seen is plotted against the wavelength or frequency of the light, the resulting curve (scotopic spectral sensitivity curve) agrees well with the absorption spectrum of human rhodopsin. If measurements on rhodopsin in solution are used in making the comparison, the agreement is close, but not exact (for discussion see Rushton, 1956b). However, the quasi-crystalline organization of visual pigments in the receptors alters slightly their absorption spectrum (see p. 26). If the hydroxylamine difference spectrum of a suspension of rhodopsin-containing particles from human rods (Wald & Brown, 1958) is taken for comparison with the scotopic spectral sensitivity curve, no significant discrepancy can be detected over the whole of that range of the spectrum (above 450 nm) in which the absorption spectrum should coincide with the hydroxylamine difference spectrum.

The threshold spectral sensitivity curve of the dark-adapted fovea (Stiles, 1939) has its maximum between 550 and 560 nm. Some colour discrimination is possible even at absolute threshold; thus the foveal threshold spectral sensitivity curve cannot be determined by a single receptive pigment. It probably depends on two or three cone pigments (see Chapter 8), and we have no means of knowing the exact manner in which it is related to the absorption spectra of these pigments, though good guesses can be made at the approximate relation.

The spectral sensitivity curve of the light-adapted extrafoveal retina was first measured at threshold by Wald (1945). The measurements were made on the 'cone' shoulder of the dark-adaptation curve (see Fig. 7.5), between the fourth and the eighth minute. The spectral sensitivity curve found by Wald was similar to that which Walters & Wright (1943) had earlier obtained by heterochromatic comparison of brightness; it resembled the foveal spectral sensitivity curve, except at short wavelengths, where it differed in a manner that could very satisfactorily be accounted for by assuming that the only relevant difference between the fovea and the 8° extra-fovea was the screening of the former by macular pigment.

The properties of regions of the retina very remote from the fovea have not been studied at threshold, but Weale (1953a) has investigated them by heterochromatic comparison of brightness. The spectral sensitivity curves have two maxima, at roughly 540 and 460 nm.

## Spatial influences on the threshold

The larger an object is, the more easily it can be seen. To define the range of conditions under which this statement is true and to assess the influence of size on visibility quantitatively has taken much careful experimenting. The information collected is at least of technical importance is helping us to predict, for situations which have not been exactly duplicated before, what tasks the visual pathway as a whole is capable of performing; and some of it can probably be used to provide evidence about how these tasks are performed.

### Relation between diameter and threshold intensity for a circular field

This was first investigated by Riccó (1877), who made careful measurements of the absolute threshold of the dark-adapted eye. No fixation point was used, but it is clear that the observations must have been extrafoveal. Riccó did not know that the fovea is the least sensitive part of the dark-adapted retina; indeed he thought it the most sensitive. He found that for circular fields of up to 42' 40" in diameter the threshold luminance of a field was inversely proportional to its area. This generalization is known as Riccó's law. Many later experimenters have confirmed it as true, at least very nearly, for the dark-adapted extrafoveal retina. The size of field for which it holds increases with increasing distance from the fovea (Hallett, Marriott & Rodger, 1962), the range being about 30' to 2°. Riccó's law may be alternatively stated as that the threshold luminous flux is independent of its spatial distribution. It represents the greatest commonly observed degree of spatial summation. A more extreme condition can be imagined, in which a given luminous flux is more effective in stimulating if spread over a large area than if concentrated. Small deviations from Riccó's law in this direction have been reported for retinal ganglion cells in the goldfish (Easter, 1968), and occasionally also in man (e.g. Fig. 2b of Glezer, 1965). With extremely small fields, Riccó's law must necessarily be true for all parts and all functional states of the retina because of the optical imperfections of the eye. Its validity for the dark-adapted extrafoveal retina, and for incremental threshold conditions where the blue-sensitive mechanism of trichromatic vision is isolated (see p. 242), is certainly of physiological and not merely of optical origin, since it holds for fields whose retinal images are unquestionably larger than that of a point source. When foveal stimuli of colours other than blue are used, the diameter of field up to which Riccó's law holds has been variously estimated at from 2′ to 6′ (Graham, Brown & Mote, 1939; Lamar, Hecht, Shlaer & Hendley, 1947; Hillman, 1958). Probably this is not quite sufficient to establish that it has more than an optical basis.

The first experimenter to investigate the relation between diameter and threshold outside the range in which Riccó's law is valid was Piper (1903), who used extrafoveal circular fields of between 2° 45′ and 26° diameter, and found that the threshold was inversely proportional to the diameter of the field, i.e. to the square root of its area. If the logarithm of the threshold intensity is plotted against the area of the field, Riccó's law corresponds to a straight line with gradient $-1$, and Piper's to a straight line with gradient $-0.5$, and if these two laws described the whole of spatial summation, all experimental points for a given region of retina under given temporal and adaptational conditions should lie on two straight lines with these gradients, intersecting sharply at a point representing the transition from one law to the other. Many later workers have investigated the relation for foveal and extrafoveal retina under a wide range of conditions. Most, though not all, have found a fairly sharp transition from a gradient of approximately $-1$ at small sizes of field to a smaller negative gradient at large; but there is no clear tendency for this smaller gradient to be equal to $-0.5$; under many conditions it has been found to be significantly less, and under a few significantly greater. Nor is the negative gradient of the part of the curve where Riccó's law is not obeyed necessarily constant; using extrafoveal retina, Graham, Brown & Mote (1939) and Barlow (1958) found it to decrease steadily with increasing size of field. Examples of observed relations between area and threshold can be seen in Fig. 7.3 (p. 172) for extrafoveal retina, and in Fig. 8.17 (p. 243) for the margin of the fovea.

As an empirical generalization, Piper's law is thus of little value, but it remains of interest as expressing the implication of two simple hypotheses about spatial summation. One of these, the hypothesis of optimal discrimination of signal from noise, attempts to describe the overall performance of the eye, and the other, that of twofold coincidences, its mode of functioning. Both will be discussed at the end of this chapter. Neither can be exactly true, since their implications, including Piper's law, are not; but they are so simple, precise, and inherently plausible that they deserve to be considered as possible starting-points for the more complex theory which will be required if all observations on the temporal, spatial and adaptational properties of the visual threshold are to be accounted for.

*Other spatial influences on the discrimination of intensities*

Lamar, Hecht, Shlaer & Hendley (1947) investigated foveal incremental thresholds with rectangular fields of various lengths and widths. They found that the threshold luminance was determined by the total perimeter of a field independently of its area, provided that its thickness exceeded about 3′.

The incremental threshold for a small test field is much higher when it lies near the boundary between two large steady fields of different luminance than when it is within either of them at a distance from the boundary. This phenomenon and its temporal counterpart are considered together on p. 185.

A technically important but physiologically complex form of discrimination is to decide which of two fields is the brighter, both being presented for a long time and the eye being allowed to wander freely from one to the other. It is common experience that the accuracy of such discriminations improves with increasing size of field, but I know of no quantitative investigation of this improvement. Illusions such as that of Fig. 6.3 show that properties of the boundary can cause systematic errors in discriminations of brightness, and suggest (as do also the experiments of Lamar *et al.*, 1947, at incremental threshold) that the improvement in accuracy with increasing size may depend on length of boundary rather than on area.

*Visual acuity*

The visual acuity is the angular resolving power of vision, or the least angular distance between two contours that can be distinguished from each other visually. Some writers prefer to use the reciprocal of the angular resolving power as the measure, so that high numbers represent good acuity. Some tests of visual acuity are not sharply distinct from tests of luminance threshold; for example detection of the position of the gap in a 'Landolt's C', i.e. an interrupted black annulus drawn on a white background, is as much a test of the luminance threshold for a small rectangular field furnished with a surround of complex shape as it is an estimate of the ability to resolve contours. The test that most clearly measures resolving power is the grating, an array of parallel rectangles alternately dark and light. Customarily these are of equal width, and at least six of each are presented. The dark rectangles need not be absolutely dark; if they are, the grating is described as having 100% contrast. The old literature on acuity is largely restricted to experiments at or near 100% contrast. Exploration of other values of the contrast (following Aubert, 1865 and Conner & Ganoung, 1935) has become fashionable since about 1960. The older literature is also largely restricted to test objects with sharp boundaries, but the fashionable test-object of the last decade is the grating with sinusoidal modulation, which was first used by Le Grand (1935). For examining the properties of optical systems, where Fourier analysis is appropriate, sinusoidal gratings have real advantages; but there is no reason to suppose that they are especially well suited for the analysis of the retinal and neural factors in vision.

The factors that limit visual acuity when light is plentiful are quite different for the fovea and for the rest of the retina. The visual acuity of extrafoveal retina is much worse than the quality of the optical image formed on it and the fineness of its pattern of rods and cones could allow (for references see Sloan, 1968). Presumably much of the spatial information contained in the responses of the rods and cones is discarded somewhere in the visual pathway, probably between the hundred million receptors and the one million optic nerve fibres.

The fovea, on the other hand, achieves a resolving power near or equal to the limits set by the fineness of its mosaic of cones and by the imperfections, due to diffraction and spherical and chromatic aberration, of its optical image. It is a striking illustration of the economical design of the eye that all these factors are so nearly limiting. With an artificial pupil of diameter less than 1·5 mm, the least resolvable angle is inversely proportional to the pupillary diameter (Cobb, 1915), as it should be if diffraction is then the sole limiting factor. Diffraction begins to affect acuity at a pupillary diameter of 2·35 mm for red light and less for light of shorter wavelengths (Shlaer, Smith & Chase, 1942). The natural pupil contracts in bright light almost but not quite into the range of sizes where resolution would be impaired because of diffraction. Chromatic aberration can limit acuity, for Shlaer, Smith & Chase (1942) and Campbell & Gubisch (1967) discovered conditions under which acuity is a little better in monochromatic light than in equally bright white light; yet the limitation is so slight that Hartridge (1947), and others to whom he refers, failed to find it. Spherical aberration can limit acuity if the pupil is dilated by means of drugs (Cobb, 1915; Campbell & Gubisch, 1967); but the natural pupil becomes large enough for the retinal image to be much blurred by spherical aberration only at very low illumination, and then acuity is not limited by the quality of the image, but by scarcity of quanta (see below). An exact theory of the best acuity that might be attained with a given mosaic of foveal cones would be difficult and probably unrewarding  An obvious idea that will not be very far from the truth is that the least resolvable distance between retinal images corresponds to the distance between adjacent cones. On this basis the foveal cone mosaic could not, for its performance, be much coarser than it is.

The best acuity that can be achieved in monochromatic light with the natural pupil is probably the same at all wavelengths (Brown, Kuhns & Adler, 1957, and the majority of other experimenters). For test gratings superimposed on backgrounds of a different colour, this does not hold. If the combination of colours is such as to isolate the red- or green-sensitive mechanism of colour vision, acuity is high; if it isolates the blue-sensitive mechanism, acuity is very much lower (Stiles, 1949; see p. 241).

The optical defects of the eye can be largely avoided by forming on the retina an interference pattern between two coherent images in or near the plane of the pupil. The interference pattern constitutes a sinusoidal grating of high contrast; the contrast is less than 100% only on account of scattering in the eye and reflexion from the pigment epithelium and choroid, and at high spatial frequencies these are probably small compared with ordinary optical aberrations. Le Grand (1935) showed that acuity is better for high-contrast interference patterns than for patterns that have passed through the ordinary optical system of the eye, though in terms of spatial frequency the improvement is small (see also Campbell & Green, 1965).

The above account of the factors limiting visual acuity has neglected the effect of luminance upon it. There is a range of rather high luminances, between about 30 and 30,000 cd m$^{-2}$, over which this is justifiable, since a large change in luminance causes only a very small change in acuity. At luminances below about 0·3 cd m$^{-2}$ (Lythgoe, 1932), acuity decreases substantially, and this may reasonably be correlated with the small number of quanta incident on (and the still smaller number absorbed in) each foveal cone: inevitable statistical fluctuations in the number of quanta absorbed become important. In these circumstances we can usefully compare the performance of the eye with that of an ideal detector of photic signals, as was first done by Rose (1948). If values are assumed for the threshold ratio of signal to root-mean-square noise and for the integrating time of the visual pathway (and estimates of these quantities are not entirely arbitrary), the luminance can be calculated at which an ideal detector with a given size of pupil would have a given angular resolving power. The ratio of this luminance to that at which the eye in fact achieves the given angular resolving power may be called the quantum efficiency of vision (see also p. 197). From the measurements of acuity made by Conner & Ganoung (1935) and others, Rose estimated that the quantum efficiency varied only over the comparatively small range from 5 to 0·5% in the million-fold range of luminance from 3 × 10$^{-4}$ to 300 cd m$^{-2}$.

Visual acuity is slightly but definitely better for vertical or horizontal lines of gratings than for oblique (Emsley, 1925). The cause is not optical, since resolution remains better at vertical or horizontal than at oblique orientation when the test-objects are sinusoidal gratings of low contrast, for which the resolvable spatial frequency is much lower and hence any optical imperfections much less important (Campbell, Kulikowski & Levinson, 1966). The mosaic of foveal cones is roughly a hexagonal lattice (Polyak, 1941), so its preferred directions, if any, would repeat every 60°, not every 90°. The superior resolving power for vertical and horizontal lines must therefore be due to properties of nerve cells and their connexions. From what is known of the receptive fields of cells of various parts of the visual pathway (see pp. 86, 103, 117), it is natural to guess that it is connexions in the striate cortex that favour the vertical and horizontal.

Adaptation to a grating of one orientation depresses the ability to resolve gratings of similar orientation and spatial frequency (Gilinsky, 1968; Blakemore & Campbell, 1968), as can be seen in Fig. 7.1. This depression, like the figural after-effects (p. 146) to which it is doubtless related, is transferred from one eye to the other, so the relevant action of the adapting stimulus must be in the brain. Again it is natural to guess that it is in the striate cortex.

## Temporal influences on the threshold

*Single flashes of light.* If a stimulus is very brief, the product of its luminance and duration (or more generally the integral of its luminance with respect to time) determines whether it is seen, and, if it is seen, the quality of sensation produced. This is Bloch's law, sometimes referred to by the name that it bears in photochemistry, the Bunsen-Roscoe law. It may be regarded as the ideal form of temporal summation, corresponding to Riccò's law of spatial summation. For stimuli of very long duration, threshold is determined by the luminance, independently of the duration. At intermediate durations there is incomplete summation. Estimates by different authors of the greatest duration at which complete summation occurs do not agree very closely; but most experimenters (Braunstein, 1923; Sperling & Joliffe, 1965, and others) have found that it is slightly greater for small fields than for large, and correspondingly that the area within which complete spatial summation occurs is slightly greater for short exposures than for long. It seems that the visual pathway can use either spatial or temporal summation in attaining threshold, but not high degrees of both at the same time. The completest published data on the interaction of spatial and temporal summation are those of Bouman (1950, 1952a) and Barlow (1958). Barlow's observations are shown in Figs. 7.2 and 7.3. The lowest set of points in Fig. 7.2 shows the effect of duration on the absolute threshold at large and at small area, and

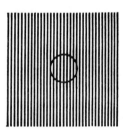

FIG. 7.1. Demonstration that adaptation to a grating specifically depresses the ability to resolve gratings of similar orientation and similar spatial frequency. Sit three metres from the page. Look for about a minute at the upper left grating, allowing the gaze to wander round the small circle in its middle. Turn the eyes to the face of the small figure in the middle. The rain in which he stands will have ceased, but will return in a few seconds. Looking at the horizontal grating or the vertical gratings of different spatial frequency does not cause the rain to disappear (from Blakemore & Campbell, 1968).

the corresponding set in Fig. 7.3 the effect of area on it at long and at short duration. Other sets of points in both figures show the effects of adaptation to a steady background, which will be discussed in the next section.

It is of some interest to establish whether there is a lower as well

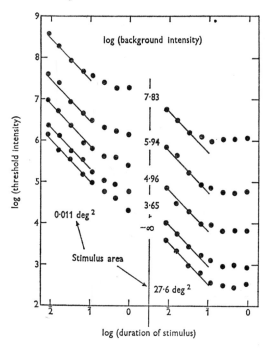

FIG. 7.2. Temporal summation at incremental and absolute threshold for a circular field centred 6° 30′ from the fixation point. Duration in seconds. The straight lines have gradient − 1. Note that the greatest duration at which complete summation occurs (gradient − 1) is slightly shorter, and the amount of summation between 0·1 and 1 sec much less, for large fields than for small (from Barlow, 1958).

as an upper bound to the duration of stimuli such that the stimulating power depends only on the total quantity of light independently of its distribution in time. For stimuli weak enough to affect only a small fraction of the molecules of receptive pigment it would be surprising on theoretical grounds if there were a lower limit. Bloch's law has been verified at threshold and for equal sensation from moderately suprathreshold flashes down to $4 \times 10^{-7}$ sec (Brindley, 1952; see also Beams, 1934, for trains of identical flashes) and it seems unlikely that it will be found to be false at shorter durations. For flashes so bright that they bleach most of the receptive pigment ($> 3 \times 10^6$ cd m$^{-2}$ sec), departures from Bloch's

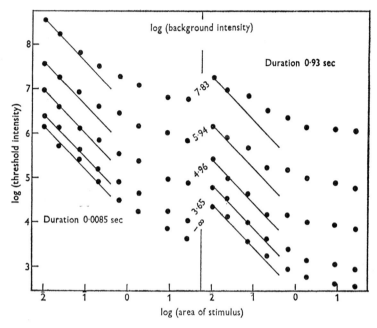

FIG. 7.3. Spatial summation at incremental and absolute threshold for a circular field centre 6° 30′ from the fixation point. Areas in square degrees. The straight lines have gradient −1 (from Barlow, 1958).

law can be detected at very short durations if the visual effects are assessed by their after-images (Brindley, 1959a). If two such flashes, each lasting 250 μsec, are presented to the same region of retina with a dark interval of only 50 μsec between, the after-image of the two together is indistinguishable from that of one of them alone; but if they are presented 4 msec apart, the after-image of the two is easily distinguished. This is an analogue of the Hagins phenomenon, and its chemical mechanism is discussed on p. 30.

*Flicker fusion frequency.* An experimentally convenient measure of the temporal resolving power of vision is the flicker fusion frequency, the least frequency of regular interruption of a photic stimulus at which no flicker is detectable. This critical frequency varies with the luminance and size of the stimulus, with the region of the retina which it illuminates, and with the state of adaptation of the eye. Its variation with luminance, the eye being adapted to the stimulus, is shown in Fig. 7.4. The break in the curve at 16 c/sec corresponds to the transition from rods to cones, as is shown by the spectral sensitivities.

It is likely that in the lower part of the curve the flicker fusion frequency may be determined by properties of the rods themselves, since the much higher frequencies resolved in cone vision suggest that the nervous pathways (which are in part shared with cones, and where not shared are anatomically rather similar) can transmit high-frequency information easily. Brindley, Du Croz & Rushton (1966) showed that the blue-sensitive cones or their pathways are in this respect (as in some others) more rod-like than cone-like; they respond to flicker only up to about 18 c/sec (see p. 244).

There is no similar reason for supposing that the fusion frequency at high luminances is determined wholly by properties of receptors themselves, in this case the red- and green-sensitive cones. The observation (Brindley, 1962a) that lights interrupted at

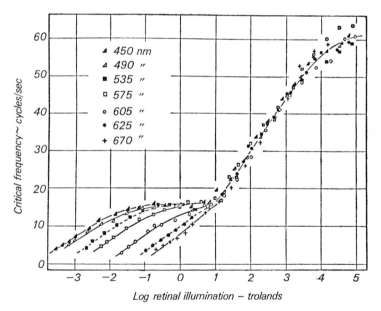

FIG. 7.4. The variation of flicker fusion frequency with retinal illumination, for a centrally fixated test field of diameter 19° (from Hecht & Shlaer, 1936).

frequencies substantially above that of flicker fusion can produce beats with alternating electric current passed through the eye at the same time proves that the photochemical mechanism of the cones is not by itself limiting. The highest frequency of inter-rupted light which when presented at high luminance (20,000 td) on a very large (40°) field, will beat with electricity is about 120 c/sec, and alternating current alone will produce a sustained sensa-tion of flicker up to at least 146 c/sec (Brindley, 1961). Yet light alone at the same high luminance on the same very large field pro-duces no sustained sensation of flicker above 90 c/sec. We must picture the visual pathway as containing two or more attenuators of high frequencies in series. The photoreceptor mechanism, up to but not including the point at which electricity acts, transmits some cyclic message at frequencies up to 120 c/sec, but much attenuated in the higher part of its range. The nervous pathway, from the point at which electricity acts, is capable, given strong enough stimuli, of transmitting some messages up to 146 c/sec. The two together can transmit to a detectable extent only up to 90 c/sec.

It is likely on grounds of economy that most of the high-frequency attenuation is in the retina, but the observation of Sherrington (1904) and Thomas (1955) that when the two eyes are stimulated together the fusion frequency is slightly higher for in-phase than for out-of-phase stimuli shows that there is some attenuation in the brain even in normal subjects. Damage to the occipital cortex can raise the flicker fusion frequency substantially in parts of the visual field where other functions are not very abnormal (Teuber, Battersby & Bender, 1960).

Ives (1922) and de Lange (1958), from the results of experi-ments on the resolution of flickering lights of various wave forms, suggested the generalization that a flicker can be resolved if and only if its first Fourier component can be resolved when all other components are replaced by steady light of the same mean lumin-ance. The proposers of the generalization would doubtless have welcomed the minor amendment (Levinson, 1960a) that a resolv-able second harmonic can substitute for a resolvable fundamental. But Levinson (1960b) showed that if the fundamental and second harmonic are both nearly resolvable, then whether their sum is resolved depends in a complex manner on the phase-relation between them. Ives's generalization is thus only roughly true.

## Adaptational influences on the threshold

Visual thresholds are influenced by the amount of background illumination that is falling on the retina when the test stimulus arrives and to which it is adapted, and also by the amount that has fallen on it in the recent past. The latter influence has attracted the more attention, but the former is easier to study in detail, since the condition in which it has to be investigated is a steady one, and we shall consider it first.

### The effect of background luminance on incremental threshold

The classical generalization about the relation of incremental threshold to background luminance (all other relevant factors remaining constant) is Weber's law (Bouguer, 1760; Weber, 1834), which asserts that they are proportional. At very low background luminances, Weber's law is obviously false. At higher luminances it is a good approximation under some circumstances and a very poor one under others. The extensive literature on the subject is well reviewed by Barlow (1957). A fair approximation to Weber's law has usually been found for stimuli of large area or long duration or both, presented on a background at least 1000 times above absolute threshold, except (Aguilar & Stiles, 1954) where the test and background wavelengths are such as to favour detection of the test stimulus by rods rather than cones and the background is brighter than 200 scotopic trolands. Barlow investigated the relation with large (4·9°) and small (5·9′) test fields placed 6° from the fixation point, exposed for short (7·6 msec) and long (940 msec) durations. He confirmed the approximate validity of Weber's law under the conditions stated above, but found that for stimuli that were both short and small and backgrounds between zero and about 10,000 times threshold, his measurements were well fitted by the equation $\Delta I = a\sqrt{(I + x)}$ which expresses the prediction of the hypothesis of optimal discrimination of signal from noise (see pp. 197–198).

The luminance of the background to which the eye is adapted influences the amount of spatial and temporal summation of which it is capable; the brighter the background, the less is the amount of summation, i.e. the less does the threshold for a large field differ from that for a small, and that for a long exposure differ from that for short (Bouman, 1952*a*; Commichau, 1955; Barlow, 1958). The same observation can be expressed in another way by saying that the longer or larger is the test stimulus, the greater is the effect of background luminance upon its threshold. Either form of the generalization can be read from Figs. 7.2 and 7.3. Reading upwards, with increasing background luminance, the curves become less steep, i.e. the effect of area or duration on the threshold decreases. Reading from left to right with increasing area of duration the vertical sets of points become further apart, i.e. the effect of background luminance on the threshold increases.

The observations of Figs. 7.2 and 7.3 were made extrafoveally under conditions where probably only rods were involved. For foveal cones, similar but smaller effects of background luminance on spatial summation have been demonstrated by Glezer (1959).

*Decrements and flashes of darkness.* Gildemeister (1914) and Bulanova and Luizov (1954) investigated the shortest interruption that can be detected visually in an otherwise continuously presented spot of light, and explored the manner in which this varies with the luminance of the continuous stimulus. The experiment may be regarded as the determination of a strength-duration curve for 'flashes' of darkness. Such 'flashes' need not be of total darkness, nor need they be brief. The detection of sudden decreases in the luminance of a $1°$ field in the $15°$ extrafoveal retina was examined by Short (1966), who found that for low background luminance the decremental threshold was regularly lower than the incremental by about a factor of two. At high background luminance the incremental and decremental thresholds are nearly equal.

Incremental and decremental stimuli can interact in a complex manner. The simplest working hypothesis for such interaction would be 'probability summation', i.e. the supposition that the probability of failing to see anything when two stimuli are presented together should be the product of the probabilities of failing to see each when they are presented separately. Experimentally this is rarely observed for stimuli presented within $\frac{1}{4}$ sec to the same region of retina. Boynton, Ikeda & Stiles (1964) and Boynton, Das & Gardiner (1966) examined the interaction of simultaneous stimuli that acted on different colour-receptive mechanisms (see p. 233). They found that in general a pair of such stimuli that were both incremental or both decremental gave less than probability summation, but an incremental stimulus to one colour-receptive mechanism and a decremental to another gave more than probability summation. Ikeda (1965) examined the interaction of increments and decrements of white light separated slightly in time, and found that a pair of incremental flashes gave more than probability summation when separated by less than about 30 msec, but less than probability summation when separated by 50 to 70 msec. An incremental and a decremental flash together gave just the reverse. Ikeda & Fujii (1966) have extended these experiments to trains of up to 10 similar flashes, and have given a theory that explains the interactions that they found. The theory could be much more stringently tested by trains of pulses whose amplitude increased and decreased smoothly during each train, and I shall be surprised if its details survive this stringent test, though the general idea behind it, which is roughly a temporal analogue of the lateral inhibition of Barlow (1953b) and Kuffler (1953) (see pp. 86–87), is very plausible.

*Thresholds in states where adaptational equilibrium has not been reached*

*Dark-adaptation: the principal facts.* The majority of investigators of non-equilibrium states have restricted themselves to the changes that occur when a person goes from bright light into total darkness. Immediately after going into the dark, he is very insensitive to dim test lights. While he remains in the dark, his sensitivity increases, for large fields by as much as 10,000 times. The process is nearly complete in 45 minutes, though a small further increase in sensitivity may occur after this. Kohlrausch (1922, 1931) found that the curve relating the logarithm of the sensitivity of extra-foveal retina to time was divisible into two sections, separated by a sudden change of gradient. The two sections were differently affected by changing the wavelength of the test light (see Fig. 7.5). During the second section of the curve, which is absent when the stimulus is placed on the fovea, sensitivity at threshold depends mainly or exclusively on rods, as is proved by the close agreement of the spectral sensitivity curve with the absorption spectrum of rhodopsin. During the first section, it has been generally supposed since Kohlrausch that cones alone determine the threshold. The

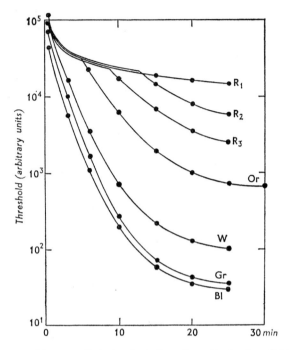

FIG. 7.5. Course of dark-adaptation for $1°$ circular fields of various colours, placed $5°$ from the fixation point. Wavebands were isolated by means of filters. $B_1$ = blue, Gr = green, W = white, Or = orange; $R_3$, $R_2$ and $R_1$ are reds of successively longer wavelength (from Kohlrausch, 1931).

supposition is supported by the finding of directional selectivity at threshold (see p. 249), and is almost certainly correct.

The increase in sensitivity found when large fields are used, as in the experiments of Fig. 7.5, is in part due to an increase in the amount of spatial summation, for Craik & Vernon (1941) and Arden & Weale (1954) found that it was much smaller with small fields (see Fig. 7.6). There is not yet any direct evidence to show whether temporal summation also increases, but the analogous findings of Bouman, Commichau and Barlow at incremental threshold suggest that it probably does.

*Dark-adaptation: evidence for a causal connexion between the regeneration of visual pigment and the increase in sensitivity.* It has

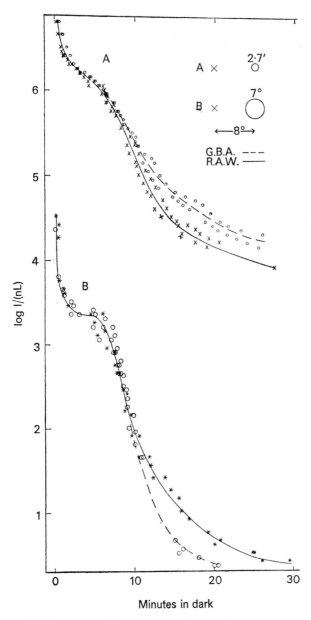

FIG. 7.6. Dark-adaptation curves for large (7°) and small (2·7) fields, 8° from the fixation point (from Arden and Weale, 1954).

been likely since the work of Ayres & Kühne (1882) that the time-courses of dark-adaptation and of the regeneration of bleached rhodopsin agree at least roughly. An impressively close correlation was established by Dowling (1960) for the rat and by Rushton (1961) for man, the logarithm of the sensitivity being linearly related to the amount of rhodopsin present.

Experiments like that of Fig. 7.6 show that the change in spatial summation during dark-adaptation follows the same time-course, at least roughly, as the change in small-field sensitivity. Thus the correlation established by Dowling and Rushton suggests that the amount of rhodopsin directly affects not only the sensitivity of the rods themselves, but also (and perhaps more importantly) the organization of the nervous pathways. The view that the correlation is not fortuitous but depends on a direct causal relation is strongly supported by the existence of similar correlations with very different time-courses for human cone-pigments (best established by Rushton & Baker, 1963) and for the rapidly regenerating rhodopsin of *Alligator mississipiensis* (Wald, Brown & Kennedy, 1957). I do not think it is appreciably weakened by the various indications (see p. 184) that the visual pigment concentration does not quite uniquely determine the threshold, i.e. that when the same visual pigment concentration is reached by two very different sequences of illumination and recovery, slightly differing thresholds can result.

*Dark-adaptation: receptor-desensitization hypotheses.* One might suppose that the regeneration of visual pigments accounted for the change in sensitivity during dark-adaptation (other than the part due to change in summation) in a simple way, the threshold being inversely proportional to the amount of rhodopsin present (or, for the early branch of the curve, the amount of cone pigment). This assumption has been widely made (e.g. Wald, 1944; Crawford, 1946); but good indirect arguments against it were provided by Lythgoe as early as 1940, and the direct measurements by Campbell and Rushton (1955) of the bleaching and regeneration of rhodopsin in the living human eye clearly disprove it. After 7 minutes, half the rhodopsin is regenerated. The threshold is then certainly not twice that of the fully dark-adapted eye, but about 500 times for large test fields and about 50 times for small.

Although the threshold, even for small fields, is not inversely proportional to the amount of rhodopsin present, we might still suppose that the loss of sensitivity when some rhodopsin is bleached depends on a desensitization of receptors of some more subtle kind. A suitable mechanism was suggested by Wald (1954): a rod might consist of many compartments, each of which would be rendered functionless by the photolysis of any one molecule of rhodopsin in it. In its original form, this suggestion agrees poorly with the quantitative relation between the threshold and the amount of rhodopsin bleached (Rushton, 1961), but by altering the number and size of compartments postulated it can be made to fit fairly well.

Any receptor-desensitization hypothesis must have great difficulty in explaining the important experiment of Rushton & Westheimer (1962), who bleached with a brief flash (to avoid eye movements) either through a grating of 2 cycles/degree or through a neutral filter that transmitted the same total amount of light, and found the rod light-adaptation produced (as measured with a uniform test field smaller than either bleaching stimulus) to be the same. Thus, apparently, bleaching some of the rods (those on which the bright bars of the grating fell) desensitizes the eye to test light that subsequently falls on their unbleached neighbours. The only way to rescue a pure receptor-desensitization hypothesis is to suppose that the retinal image of Rushton and Westheimer's grating was very much less contrasty than they suppose. There is perhaps room for a little scepticism of this kind, but not enough to save pure receptor desensitization.

*Dark-adaptation: bleaching-signal hypotheses.* A quite different notion about the nature of dark adaptation is that when rods are partly bleached (say up to 20%), they continue to send their normal signals with nearly normal sensitivity when light falls on them, but send in addition, whether light is present or not, signals that depend on the fact that some of their rhodopsin is absent from its proper sites.

An attractive hypothesis of this kind is that of Barlow (1964), who suggested that partly bleached receptors give signals indistinguishable from those provoked by light, and that the changes of retinal organization associated with light-adaptation depend wholly and immediately on these signals. The hypothesis at once explains the similar time-course of dark-adaptational changes in spatial

G

summation and small-field threshold, and the similar effects on spatial summation of light-adaptation and background (see especially Blakemore & Rushton, 1965), both of which, on receptor-desensitization hypotheses, are mere coincidences. It provides a satisfactory basis for the lasting constriction of the pupil produced by light adaptation in the absence of lasting light (Alpern & Campbell, 1963), and is consistent with the observations of Rushton and Westheimer on bleaching through gratings. Perhaps its strongest support comes from the comparisons by Barlow & Sparrock (1964) of the brightnesses of positive after-images and stabilized retinal images; to account for light-adaptation, the signals from bleached receptors need to correspond to quite large amounts of background light. One might have supposed, from the inconspicuousness of positive after-images in everyday life, that they were not as bright as Barlow's hypothesis demands; but much of their inconspicuousness comes from the fact that they are stabilized on the retina, and Barlow and Sparrock found that when backgrounds that have the same threshold-raising effect are stabilized on the retina by contact-lens optics and compared with them, they match.

There are, however, some facts that are probably inconsistent with Barlow's hypothesis in its pure form. Rushton (1965c) found different kinds of spatial integration for the effects of background and of bleaching on the threshold for a uniform test light; however, eye movements, which could easily be very important, were not controlled, so the conflict with Barlow's hypothesis is not sharp. Westheimer (1968a) found a difference between the spatial effects of bleaching and background which cannot be attributed to eye movements. It was already known (Westheimer, 1965) that the raising of threshold by a steady background could be partly counteracted by providing a bright surround to the field. Teller, Andrews & Barlow (1966) showed that this phenomenon was independent of eye movements. Westheimer (1968a) then showed that it was absent if the steady background or the surround or both consisted of bleached receptors instead of real light of similar threshold-raising capacity. Another experiment in which light-adaptation and background behave differently is that of Ernst (1968), who found that when a given degree of light-adaptation and a given superimposed background are alike in the degree to which they raise the threshold for a flash, the background permits the better resolution of superimposed flicker.

Some electrophysiological observations also are difficult to reconcile with the pure form of Barlow's hypothesis; in particular, Naka & Rushton (1968) found that the effects of bleaching and of background on non-colour-sensitive s-potentials in fish are very different.

To explain the differences in spatial integration for background and for bleaching that he found, Rushton (1965c) proposed a new bleaching-signal hypothesis. He suggested that the signals produced by bleaching were of a different kind from those produced immediately by the absorption of light, and were proportional to the amount of pigment bleached, where those of Barlow's theory had to be an exponential function of the amount bleached. Rushton's suggestion explains his own observations and those of Westheimer (1968a), but on grounds of biological economy it is as unattractive a notion as Barlow's is beautiful. Long-lasting insensitivity after quite small degrees of bleaching is a wholly disadvantageous property of the visual system. Barlow attributes it to a regrettable but unavoidable feature of the organization of rods and cones; Rushton to an inessential feature, fairly complex and altogether harmful. One would have expected such a feature never to have evolved, or to have been eliminated by natural selection if it did evolve.

*Dark-adaptation: mixed hypotheses.* The experimental objections to pure receptor-desensitization hypotheses and to Barlow's pure bleaching-signal hypothesis can probably be overcome by mixing them, i.e. by supposing that when receptors are partly bleached they become both insensitive and noisy, and that the insensitivity of the eye as a whole is due partly to insensitivity of receptors and partly to their noisiness. Such mixed hypotheses are untidy when one wishes to test them, since they contain many undetermined parameters, but, unlike Rushton's hypothesis, they do not conflict with biological economy.

*Dark-adaptation: automatic gain control.* On any bleaching signal hypothesis of the nature of light-adaptation, whether the signal produced by absence of visual pigment from its proper place is distinguishable (Rushton) or indistinguishable (Barlow) from that produced immediately by light, some kind of automatic gain control is needed to vary the sensitivity of postreceptoral stages of the visual pathway according to the strength of the bleaching signals. Rushton (1965c) has made detailed suggestions about how the automatic gain control might work, but such suggestions are difficult, perhaps impossible, to test by psychophysical experiments.

*Dark-adaptation: is the state determined by the amount of pigment bleached?* All the theories that we have discussed assume that the amount of visual pigment bleached uniquely determines the adaptational state. The experimental evidence clearly shows that this is roughly true, but contains several indications that it is not exactly so. Crawford (1946), in a very thorough investigation, found that the whole course of dark-adaptation for a 30' extrafoveal field, excluding the first 20 seconds, was unaffected by varying the time during which a given quantity of adapting light was delivered, provided that this did not exceed 90 seconds. Crawford's findings are thus consistent with the view that, except for the first 20 seconds, all the adaptational change in sensitivity is determined by the amount of rhodopsin present; however, only rather small degrees of bleaching were investigated. With a larger test field, 2°, and larger amounts of bleaching Mote & Forbes (1957) found that the adapting effect of a given quantity of light was *not* independent of its temporal distribution. Rushton (1963d) showed that a maximal brief flash (msec) applied to the human eye bleaches only about half as much rhodopsin as does a maximal exposure lasting several seconds (Hagins phenomenon, see p. 30). Yet the course of rod dark-adaptation is the same after both kinds of bleaching. Rod thresholds can be measured from 16 minutes after the bleaching, by which time regeneration is too nearly complete for opthalmoscopic densitometry to measure its incompleteness; however, extrapolation of the regeneration curves strongly suggests that identical thresholds are being obtained with different amounts of rhodopsin. For cone pigments, where threshold measurements and ophthalmoscopic densitometry can satisfactorily be applied at the same time, Rushton & Baker (1963) found a clearer and more extreme discrepancy of the same kind: a maximal one-millisecond flash bleaches only about 60% of the foveal cone pigments, and their regeneration, measured by ophthalmoscopic densitometry, proceeds substantially *faster* than after a maximal 10-second bleach. Yet the corresponding foveal dark-adaptation proceeds not faster but slower, the threshold at 4 minutes being some 3 times higher.

Rushton and Baker call the raised threshold after the maximal one-millisecond flash the '$\theta$ effect', and show that it occurs only when a substantial fraction of the visual pigment molecules absorb more than one quantum within a short time.

Comparison of foveal dark-adaptation curves after total bleaches lasting 5 seconds or 2 minutes (Baker, Fulton & Rushton, 1969) reveals a distinction whose significance is more immediately clear. Though these both bleach all the pigment, regeneration as measured by ophthalmoscopic densitometry proceeds initially twice as fast after the briefer bleach. In the corresponding dark-adaptation curves, equal thresholds are found at equal *degrees* of regeneration, not when the *speeds* of regeneration are equal.

The rapid changes in sensitivity that occur within the first half-second of extinguishing an adapting light (Bouman, 1952$b$) clearly cannot depend on corresponding changes in the amount of visual pigment present.

*Incremental thresholds near a spatial or temporal discontinuity in the background.* If an incremental stimulus is presented near the boundary between two uniform fields (Blachowski, 1913; Fiorentini, Jeanne and Toraldo di Francia, 1955) or flashed at at a time close to that at which the luminance of the background is suddenly altered (Crawford, 1937, 1947; Baker, 1955; Hallett, 1969$a$), the threshold for it is raised more than it is when it is presented centrally in either uniform field or on either of the steady luminances between which the sudden alteration occurs. *Temporal* effects of this and related kinds are easy to demonstrate unequivocally. For the *spatial* effects, simple experiments done without stabilization of the retinal image need to be checked with stabilization before the possibility that the relevant discontinuities are temporal or spatio-temporal rather than purely spatial can be firmly excluded. When this control is done, some spatial effects remain (Teller, Andrews & Barlow, 1966), but others disappear or are greatly reduced (Teller, 1968).

Spatial influences on incremental thresholds cannot be interpreted as effects of boundary alone, without considering what the boundary separates; Westheimer (1967$a$) showed that a pair of concentric narrow black and white rings on an otherwise uniform grey background do not affect the photopic incremental threshold at their centre, though a boundary of similar conspicuousness that separates different luminances has a large effect. It may, as West-heimer argues, be more profitable to interpret the spatial influences in terms of a lateral inhibition similar to, and doubtless largely dependent on, that which is well known for retinal ganglion cells.

## Thresholds during eye-movements

We see, essentially, in a series of snapshots. Our eyes remain, for half a second or so, pointing at a fixed point in space, stabilized against head movements by the vestibulo-ocular reflexes. This is a 'fixational pause'. Then the eyes move quickly through an angle that may vary from $\frac{1}{10}°$ to over 90°. This is a 'saccadic movement', and ends in another fixational pause. The stabilization of the eyes during a fixational pause is imperfect, but there is no good ground for supposing the imperfection to be beneficial; it seems rather that the oculomotor mechanisms perform the difficult task of keeping the eye still just, but only just, as well as is needed.

During a saccadic movement the whole retinal image moves. Such a movement of the retinal image, if it occurred during a fixational pause, would be very conspicuous, but when produced by the movement of the eyes themselves it is inconspicuous. This simple fact suggests that messages from the retina are in some sense suppressed or inhibited during saccadic eye movements. Latour (1962) and Volkmann (1962) showed that during large voluntary saccadic eye movements the threshold for a very brief flash of light is raised. Beeler (1967) found a similar though smaller rise in threshold during involuntary saccadic movements, which are commonly of smaller amplitude than the voluntary ones studied by Latour and Volkmann.

Events in the brain which may perhaps be related to the sup-pression during saccadic movements are considered on p. 103.

## The nature of visual thresholds

### One quantum suffices to stimulate a rod

Hecht, Shlaer & Pirenne (1942) measured the absolute threshold for short flashes of light of wavelength 510 nm. They used a small extrafoveal field (10′ diameter) which, according to their estimate, illuminated about 500 rods. They found that a flash of 1 msec duration was seen on 60% of presentations if it contained from 54 to 148 quanta, as measured at the cornea. Estimates of the amount of light lost by reflection and absorption in the eye before it reached the retina and the fraction of light incident on the retina which is absorbed in rhodopsin lead to the conclusion that even from the brightest of these flashes (148 quanta) not more than 14 quanta on the average could be absorbed by rhodopsin. The probability that from such a flash at least one of the 500 rods receives two or more quanta is

$$1 - e^{-14}\left(1 + \frac{14}{500}\right)^{500} = 0\cdot178,$$

i.e. substantially less than 0·60, the probability of seeing the flash. Hence the absorption of two quanta in the same rod cannot be necessary for seeing; one quantum must suffice to stimulate a rod.

If the data of Hecht, Shlaer and Pirenne alone are considered, it is possible to question this conclusion by doubting whether the margin between 0·178 and 0·60 is sufficient to cover the uncertainties of the estimates for the fraction of incident light which is absorbed in rhodopsin and the number of rods illuminated by the retinal image of a 10′ field (350 would be a more likely figure than 500). However, if the same argument is applied to Stiles's measurements (1939) of the threshold for large fields, it establishes the sufficiency of one quantum beyond all possible doubt. Stiles found that a square of extrafoveal stimulus of side 1·04° centred 5° from the fixation point and of duration 63 msec had to contain 122 quanta of wavelength 510 nm, as measured at the cornea, in order to be seen on 50% of presentations. Such a field illuminates at least 12,000 rods. Applying the correction of Ludvigh & McCarthy (1938) for transmission losses, 62 of the 122 quanta incident on the cornea reach the retina. Even if we make the impossible assumption that *all* of these are absorbed in rhodopsin, the probability that at least one rod receives two or more quanta is

$$1 - e^{-62}\left(1 + \frac{62}{12000}\right)^{12000} = 0\cdot162,$$

i.e. substantially less than 0·5.

We do not yet know what kind of message is transmitted by a rod when it receives a single quantum. But whatever its nature, it is not only occasionally that the absorption of a quantum causes it to be transmitted; the arguments of Hecht, Shlaer and Pirenne quoted on pp. 189–191 prove that at least a third of such absorptions, and perhaps all of them, are signalled to the neural layers of the retina and can influence the kind of message that they relay to the brain.

## How many quanta are required to stimulate a foveal cone?

Stiles (1939) found that for a square foveal stimulus of side 1·04° and duration 63 msec, 17,900 quanta of wavelength 580 nm, as measured at the cornea, were required in order to be seen on 50% of presentations. Such a field illuminates about 9500 foveal cones. Applying the appropriate correction from Ludvigh and McCarthy, 59·4% of the corneal quanta, i.e. 10,600, arrive at the retina. There is no good evidence by which we can estimate what fraction of these are absorbed in receptive pigments. To obtain an upper bound for the number of quanta required to stimulate a cone, we may assume that all are absorbed, and proceed as follows:

Let the mean number of quanta absorbed per cone be $I$, the total number of cones illuminated by $N$, the probability that $m$ or more quanta will be absorbed in a given cone be $P$, and the probability that $m$ or more quanta will be absorbed in at least one of the $N$ cones be $Q$. Then if the actual numbers of quanta absorbed follow a Poisson distribution,

$$P = 1 - e^{-I}(1 + I + I^2/2! + I^3/3! + \ldots \ldots \ldots \text{to } m \text{ terms}),$$

and
$$Q = 1 - (1 - P)^N$$
$$= 1 - e^{-IN}(1 + I + I^2/2! + I^3/3! + \ldots \ldots \ldots \text{to } m \text{ terms})^N.$$

For the experimental condition, $N = 9500$, and $IN = 10,600$. We require the greatest value of $m$ such that $Q \geqslant 0·5$. Substituting the numerical values, we find that $m = 7$ gives $Q = 0·786$, and $m = 8$ gives $Q = 0·190$. Hence $m$ cannot exceed 7.

If Stiles had used stimuli of very short duration, say 1 or 10 msec, instead of 63 msec, he would probably have obtained a threshold lower by a factor of about 1·2 (correction derived from the data of Bouman & van der Velden, 1948). A more realistic, though uncertain, estimate of the fraction of light incident on the retina which is absorbed in receptive pigments might be $\frac{1}{2}$. If on these grounds we diminish our estimate of $IN$ by a factor of 2·4, the greatest value of $m$ giving $Q \geqslant 0.5$ becomes five instead of seven. We cannot expect that new experimental evidence on the fraction of incident light absorbed will greatly reduce the upper bound for $m$ set by this argument. For example, if we take the improbably low value of $\frac{1}{5}$ for the fraction of light at the retina absorbed by the receptive pigments of foveal cones, so that $IN$ is only $10,600/5 \times 1.2 = 1767$, the upper bound for $m$ becomes 3.

There are no arguments from sensory experiments which establish a lower bound for the number of quanta required to excite a cone, except that this obviously cannot be less than 1; thus we can say only that the number of quanta which must be absorbed within one short period in order to stimulate a foveal cone is certainly between 1 and 7 and probably between 1 and 5.

## The condition for absolute threshold for a small and brief stimulus

Hecht, Shlaer & Pirenne (1942) estimated by two independent methods the number of quanta which must be absorbed in rhodopsin, during one millisecond and within a circular extrafoveal field of 10° diameter, for a flash of light to be detected. One of the methods was outlined above; the amount of light was measured physically, and estimates were obtained for the fraction lost by reflexion and absorption before reaching the retina, and the fraction of that light which reached the retina which was absorbed in rhodopsin. The agreement between the scotopic sensitivity curve of the eye and the absorption spectrum of rhodopsin showed that the latter fraction could not exceed $\frac{1}{5}$, and the two corrections together showed that the number of quanta which needed to be absorbed for the flash to be seen on 60% of presentations was not more than 14.

The second method used by Hecht, Shlaer and Pirenne does not directly estimate the number of quanta absorbed in rhodopsin, but the number which must contribute information to account for the way in which the probability of seeing a flash varies with its physical intensity. The authors first considered the ideal hypothesis that the flash is always seen if the number of quanta absorbed from it equals or exceeds a small integer $m$, and is never seen if the number absorbed is less than $m$. As in the argument for cones on p. 188 if the actual numbers of quanta absorbed follow a Poisson distribution, the probability that $m$ or more will be absorbed, and therefore that the flash will be seen, is

$$P = 1 - e^{-I}(1 + I + I^2/2! + I^3/3! + \ldots\ldots\ldots \text{ to } m \text{ terms}),$$

where $I$ is now the mean number of quanta absorbed from a flash, averaged over many flashes of the same physical intensity. $I$ is proportional to the physical intensity, so that the curve obtained by plotting $P$ against $\log I$ is directly comparable with that obtained by plotting the experimentally determined frequency of detecting a given flash against the logarithm of its physical intensity. Hecht, Shlaer and Pirenne found that the experimental curves were very similar to the theoretical ones for values of $m$ about six, the values giving the best fit being 6, 7 and 5 for the three authors respectively. If to the ideal detector postulated by Hecht, Shlaer and Pirenne we add 'noise' events indistinguishable from the absorption of quanta, or biological variations causing the probability of seeing to be less than unity when the $m$-quantum condition is met, then it can be proved that the gradient of the curve relating the frequency of seeing to the logarithm of the physical intensity must at every value of the frequency of seeing be less than the corresponding gradient for the same value of $m$ without the disturbing factor (Brindley, 1954$b$; Pirenne & Marriott, 1955). Since the gradient increases with $m$, agreement of an experimental curve with a theoretical one for a given value, say $m'$, of $m$ implies, if seeing depends basically on an ideal detector of the kind postulated by Hecht, Shlaer and Pirenne, that this detector requires $m'$ quanta *or more*. It would not, however, be quite accurate to say that the results of Hecht, Shlaer and Pirenne prove that five or more quanta must always be absorbed for a small and short stimulus to be detected; the hypothesis, for example, that the probability of seeing is 1 if five or more are absorbed, 0·1 if four are absorbed, and 0·01 if three are absorbed is biologically not very unlikely, and its observable implications are scarcely distinguishable from those of a rigid five-quantum threshold, though it involves that occasionally as few as three quanta will be detected. The simplest firm statement that can be made is that the *mean* number of quanta absorbed in rhodopsin and successful in stimulating rods when a small and brief stimulus is seen (and *a fortiori* for a large or long one) is equal to or greater than five.

Barlow (1956) has developed a unified theory of absolute and incremental thresholds according to which an important factor limiting the absolute sensitivity of the eye is the occurrence of events, perhaps thermal decompositions of rhodopsin molecules, which the retina cannot distinguish from the absorption of quanta in the rods. If this theory is correct, the number of quanta required for threshold may be substantially greater than five. Barlow finds that the frequency-of-seeing curves of Hecht, Shlaer and Pirenne are well fitted by the specific assumption that a coincidence of twenty-one quantum-like events is required for threshold, nine of these on the average being contributed by 'noise' and eleven by the flash of light. The theory gives reasonable values for the number of false positives found in threshold experiments, and explains how different frequency-of-seeing curves can be obtained according to whether the subject sets himself a high or low standard of certainty that he has seen the flash.

A point that is clear from the argument of Hecht, Shlaer and Pirenne, whether or not Barlow's amendment of it is correct, is that the number of quanta that must contribute information before a short and small flash is seen cannot be much less than the total number of quanta absorbed. Every, or nearly every, quantum that acts on a molecule of rhodopsin, whatever may be the position of that molecule in relation to the surface of the rod in which it lies, must cause that rod to transmit a message to the neural layers of the retina which they can use in 'deciding' whether a flash of light has been received.

In 1944, isolated in Holland during the war and unaware of the work of Hecht, Shlaer and Pirenne, H. van der Velden published a paper in which he presented three arguments in favour of a theory, part of which was that the absorption of two quanta sufficed to allow the detection of a small and short stimulus. Two of van der Velden's three arguments depended on temporal and spatial summation. They are important arguments, and will be considered further on pp. 193–195, but they are not compelling evidence for a two-quantum condition for the small-field threshold. The third argument is the same as that of Hecht, Shlaer and Pirenne from frequency-of-seeing curves. Van der Velden's frequency-of-seeing curve fits the Poisson sum for $m = 3$ and is clearly steeper than that for $m = 2$, but since the evidence from spatial and temporal summation seemed to him strong, he accepted it as consistent with a two-quantum hypothesis. In fact the evidence from spatial and temporal summation is weak, and van der Velden's frequency-of-seeing curve disproves the very hypothesis that it was taken to support. It is not inconsistent with a five-, six- or seven-quantum condition for threshold, because the argument from frequency-of-seeing curves necessarily gives a lower bound for the number of quanta required.

### Thresholds for large and long stimuli

The laws of Riccó and Bloch imply that the number of quanta required for threshold is independent of duration and area provided that these do not exceed certain critical values. For longer times and larger areas, the number of quanta which must fall on the retina, and hence the number which must be absorbed in rhodopsin, increases. The neural mechanism, whatever it may be, which correlates single-quantum events occurring in different rods and allows a message to reach consciousness when some small number, perhaps five or six, occur together, is evidently not indifferent to where and when the events occur; it functions best, or perhaps only functions, when they are close together in time and space. It is of interest to suggest some simple hypotheses which describe what the neural mechanism may do in terms of the scoring of coincidences between single-quantum events. These hypotheses have implications concerning frequency-of-seeing curves, and the relation of threshold to duration and area, which may readily be compared with experimental results.

*The hypothesis of independent sensitive units.* Van der Velden (1944) suggested that the extrafoveal retina was organized in the form of a large number of similar overlapping 'sensitive units', each containing some hundreds of rods, and each independent of its neighbours, in the sense that if a certain quantal condition was met within any one unit a flash of light was seen, and if it was met in none of them, no flash was seen. According to van der Velden, the appropriate quantal condition was the absorption of two quanta; but the hypothesis of 'sensitive units' is best considered independently of any specific assumptions about what the units may do.

The hypothesis implies that for any two non-contiguous stimuli, and to a close approximation for two contiguous stimuli if they are large compared with a sensitive unit, the probability of seeing neither when both are presented together is equal to the product of the probabilities of failing to see each when presented separately.

For extremely large stimuli (Pirenne, 1948), and for stimuli falling on widely separated areas of the retina (Denton & Pirenne, 1951), this implication has been tested and found to be true. For contiguous stimuli of moderate size no direct attempt to test it has been reported, but relations between frequency-of-seeing curves and area-threshold curves can be deduced from it which can be compared with published data. Let $Q_1$ and $Q_2$ be the probabilities of seeing a flash of a given luminance and duration, presented to regions of the retina of area $A_1$ and $A_2$ respectively. Then if the part of the retina concerned is homogeneous, and $A_1$ and $A_2$ large compared with a sensitive unit,

$$(1 - Q_1)^{A_2} = (1 - Q_2)^{A_1}.$$

This relation imposes a restriction on admissible frequency-of-seeing curves which is independent of any assumptions about what the units may do, provided only that they are independent according to the definition given above. In particular we can prove from it (see Brindley, 1959$b$) that if the relation of threshold to area at probability-criterion $Q_0$ is of the form $AI^n = c$, where $n$ and $c$ are constants, the frequency-of-seeing function must be of the form $Q = 1 - e^{-kI^n}$ where $K$ is the positive constant $Ac^{-1} \ln \dfrac{1}{1 - Q_0}$.

Temporal summation can be treated in the same way. If we suppose that a flash is seen if and only if a certain quantal condition is met within an interval $\tau$, then all time-segments of duration $\tau$ within the flash are temporal analogues of van der Velden's 'sensitive units' in space. If the flash is long compared with $\tau$, a relation of threshold to duration of the form $tI^n = c$ implies a frequency-of-seeing function of the form $Q = 1 - e^{-kI^n}$ where $k$ is the positive constant $tc^{-1} \ln \dfrac{1}{1 - Q_0}$.

The deductions are directly applicable to the data of Bouman & van der Velden (1947), who measured the relation of absolute threshold to area and duration at probability-criterion 0·6, and the frequency-of-seeing functions at all combinations of four durations and four areas. Bouman and van der Velden found that for small areas the relation of threshold to duration was of the form $tI^2 = c$ in the range from 100 msec to 5 sec. The corresponding frequency-of-seeing curves, as can be seen in Fig. 7.7, fit the theoretical relation $Q = 1 - e^{-kI^2}$ very well; the data are consistent with the hypothesis of temporal independence. The relation of threshold to area at short durations was found to be of the form $AI^2 = c$ (Piper's law) in the range from $37'$ to $1390'$ diameter. Here the corresponding frequency-of-seeing curves (Fig. 7.7) obviously do

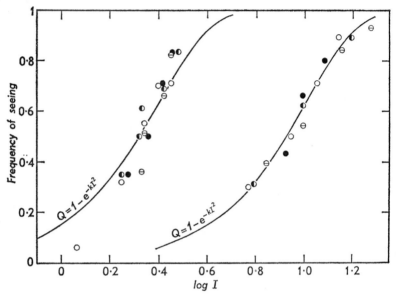

FIG. 7.7. The frequency-of-seeing curves of Bouman & van der Velden (1947) for large area and short duration (*left*) and for small area and long duration (*right*). The latter are in good agreement with the theoretical relation $Q = 1 - e^{-kI^2}$; the former clearly are not.

not fit the theoretical relation $Q = 1 - e^{-kI^2}$; the hypothesis of spatially independent units small in comparison with the sizes of field used is clearly contradicted. The same consistency of the data with the hypothesis of temporal independence and inconsistency with that of spatial independence appears if we select for analysis the area-threshold relation at long duration and the duration-threshold relation at large area. For absolute thresholds in general it appears that the detection of stimuli longer than the greatest duration at which the Bunsen-Roscoe law holds (say $T$) can be accounted for, at least nearly, by the hypothesis that a fixed quantal condition must be met within a time-segment of duration $\tau$ approximately equal to $T$. The corresponding spatial condition is not valid, some interaction between different regions of the retina in attaining threshold occurring up to areas at least ten and perhaps a hundred times greater than the largest at which Riccó's law holds.

*The simple two-quantum hypothesis.* Van der Velden's simple hypothesis, that the absorption of two quanta within a time-segment $\tau$ in one 'sensitive unit' is necessary and sufficient for vision, predicts that for areas exceeding about twice that up to which Riccó's law holds, and for durations exceeding about twice that up to which Bloch's law holds, the threshold luminance will be related to area and duration by $AI^2 = c_1$ and $tI^2 = c_2$. These relations are obeyed closely by the experimental data of van der Velden and approximately by those of other experimenters. We have seen that the simple two-quantum hypothesis is clearly contradicted by the frequency-of-seeing curves obtained with small and short stimuli. For the frequency-of-seeing functions for large and long stimuli, it implies that they can in no case have, at any value of $Q$, a gradient greater than that of $Q = 1 - e^{-kI^2}$. A proof of this is given in Brindley (1954$b$), and generalized to coincidences of any order in Brindley (1963$b$). As can be seen from Fig. 7.7, in some cases the frequency-of-seeing curves have experimentally a much greater gradient, and provide an alternative clear disproof of the hypothesis. If, then, the simple two-quantum hypothesis were the only possible explanation of the observed laws of temporal and spatial summation, we should have an irresolvable contradiction. In fact it is not the only possible explanation, and we may reject it without disquiet.

*Baumgardt's modification of the two-quantum hypothesis.* Baumgardt (1953) suggested that the condition for detection of a flash of light was not the absorption of two quanta within one sensitive unit, but the simultaneous or nearly simultaneous occurrence of several such two-quantum coincidences in different sensitive units. The laws of temporal and spatial summation implied by this hypothesis are almost the same as those implied by the simple two-quantum hypothesis, and if the number of two-quantum coincidences required and the spatial and temporal bounds within which they must occur are suitably chosen, it can be made to account satisfactorily for observed frequency-of-seeing curves. It stands, therefore, as a tolerably simple hypothesis which has not yet been disproved, and which accounts for a wide range of observations. Nevertheless, it seems unlikely that it is a close approach to the truth, because Bouman (1952b) and Barlow (1958) have shown that the laws of temporal and spatial summation are almost the same when stimuli are presented on a background sufficient to raise the threshold tenfold as when they are presented on zero background. The number of quanta absorbed from these incremental stimuli must be very nearly ten times the number absorbed at absolute threshold, so that detection cannot then depend on the scoring of twofold coincidences. It is not very plausible to suppose that the same law of spatial and temporal summation depends on one mechanism at absolute threshold, but on an entirely different mechanism at incremental threshold.

Baumgardt's theory deserves to be remembered and thought about, but as a possible starting-point for a more complete theory of how the visual pathway filters the information given to it by the rods and cones, and determines what part of it shall be allowed to reach consciousness, rather than as likely to be the precise mechanism of absolute threshold.

*The hypothesis of optimal discrimination of signal from noise.* A very different approach to the problem of the nature of visual thresholds is to defer considering the probable mechanisms involved, and to concentrate attention on the fundamental task that the retina and visual pathway have to perform, namely the extraction from the optical image of as much information as is allowed by the unavoidable statistical fluctuations in the number of quanta that can be absorbed from it, and by comparable fluctuations in the number of events such as the thermal decomposition of molecules of rhodopsin, which may be indistinguishable from the absorption of quanta. This approach was first used by Rose (1948), who pointed out that the performance of an ideal detector of light signals of intensity $\Delta I$ and duration $t$ subtending a solid angle $A$ and presented upon a background of intensity $I$ is given by

$$\Delta I^2 = \frac{h^2 I}{AtS}$$

where $h$ is the ratio of signal to root-mean-square statistical fluctuations which the detector accepts as threshold, and $S$ the area of its receiving surface. The quantities must be measured in corresponding units; for example if $t$ is expressed in seconds, $A$ in steradians and $S$ in square centimetres, $I$ and $\Delta I$ must be in quanta sec$^{-1}$ cm$^{-2}$ steradian$^{-1}$.

To describe the performance of real detectors, a number $\theta$ must be placed in the denominator to account for the factor by which their performance falls short of that of the ideal detector, the real detector performing as well as an ideal one which makes use only of a fraction $\theta$ of the quanta incident upon it. $\theta$ is called by Rose the quantum efficiency. Obviously it cannot exceed the fraction of incident quanta actually used by the real detector.

Estimates of the quantum efficiency of the human eye as a detector that were made by Rose involve two uncertainties. The first is the value of $h$, which could in principle be determined from the number of errors made by the detector, but was not estimated by Rose. The second comes from the practice of taking for $t$ not the duration of the stimulus, but the supposed integrating time of the eye, the stimulus actually used being of very long duration. Estimates that are free from these two uncertainties have been made by Barlow (1962), who found that for the discrimination of two small near-threshold extrafoveal stimuli, superimposed on a large background, the quantum efficiency is about 0·05 when the background is very weak, begins to fall when the background reaches about $10^{-3}$ scotopic trolands, and is about 0·001 on a background of 2 scotopic trolands. By a similarly satisfactory method, Hallett (1969$b$) obtained quantum efficiencies at absolute threshold around 0·1. The quantum efficiencies found by Hallet are so high that they suggest, and may perhaps suffice to prove, that the quanta that are effective in exciting rods are not merely all those that cause bleaching of rhodopsin molecules but all those that are absorbed in rhodopsin molecules. This is probably about twice as great (see pp. 12–14).

A hypothesis that has been developed by several writers, and is most clearly expressed by Barlow (1957), is that the eye tends evolutionarily to achieve, and in fact comes close to achieving, a high constant quantum efficiency under a wide range of conditions, limited only by the fraction of incident quanta actually absorbed in visual pigments. If assumptions are made about the nature of the

noise from which the signal has to be discriminated, this hypothesis has implications for the relation of incremental thresholds
not only to background luminance but also to area and duration. If
the relevant noise is assumed to be independent of the area and
duration of the stimulus, as is reasonable if the stimulus is small in
comparison with the region over which the system integrates
noise, the threshold should be inversely proportional to area and
duration (Riccó's and Bloch's laws). If the relevant noise is
assumed to be proportional to the area and duration of the
stimulus, as is reasonable for large and long stimuli, the threshold
should be inversely proportional to the square roots of area and
duration (Piper's law and its temporal counterpart). These implications are the same as those of the two-quantum hypothesis. Since
they are not exactly true, the hypothesis of optimal discrimination
of signal from noise cannot be exactly correct; but their approximate truth adds to the plausibility of supposing that optimal
discrimination of signals from noise, partly of retinal origin but
mainly arising from statistical fluctuations in the number of quanta
absorbed, is one of the principal requirements towards which
evolutionary pressure tends to make the eye evolve, subject to the
limitations of its materials and those imposed by the need for
compromise with other desirable qualities.

# 8

## COLOUR VISION

### The trichromacy of colour discrimination

A CONSPICUOUS difference between the discrimination of qualities by the human eye and by some other sense organs is the eye's
limitation, under most conditions, to three degrees of freedom.
For the senses of hearing and smell, the number of distinguishable
qualities of sensation is immensely large, and it is not generally
possible to imitate one odour or sound exactly by a mixture of a
small number of different ones. For the eye, however, the number
of distinguishable qualities (excluding spatial and temporal variations), is very much smaller, and there are simple binary mixtures,
spectroscopically very different from each other, between which no
difference can be detected visually. Given any four lights, whether
spectroscopically pure or not, it is always possible to place two of
them in one half of a foveal photometric field and two in the other
half, or else three in one half and one in the other, and by adjusting
the intensities of three of the four lights to make the two halves of
the field appear indistinguishable to the eye. This property of
human colour discrimination, whereby the adjustment of three
independent continuous controls makes possible an exact match,
though two are generally insufficient, is known as trichromacy.

To the extent that trichromacy is true for the mixing of lights, it
must be true also for the mixing of pigments which selectively
absorb light, provided that they do not interact chemically. It is
approximately true also in the simpler form that given three lights,
for which the best choice is red, green and blue, or three pigments,
for which the best choice is magenta, blue-green and yellow
(though red, blue and yellow will do fairly well), it is possible by
mixing them to match any other colour. The means for discovering
trichromacy have therefore been available to many men for
many centuries, and the history of its discovery as an empirical

generalization, and of the later appreciation that it was a property of the eye and not of light or of coloured materials, is curious and interesting.

### History of the discovery of trichromacy

Leonardo da Vinci (1452–1519) in his *Trattato della Pittura*, first published posthumously in 1651, has much to say about pigments, but surprisingly little about the results of mixing them. He recognizes six 'simple colours', white, yellow, green, blue, red and black, but says that white and black are not accepted as colours by all philosophers. Later he casts doubt on this classification, and suggests that blue and green are not truly simple, because blue is compounded 'of light and of darkness', and green of blue and yellow. The production of green by mixing or superimposing blue and yellow pigments is discussed at length by Robert Boyle in his book *Experiments and Considerations touching Colours* (1664). Boyle does not mention any other mixtures. Mariotte, writing in 1681, is more thorough. Of colours, he states: 'Il y en a a cinq principales; le blanc, le noir, le rouge, le jaune, et le bleu. Toutes les autres se peuvent faire par le mélange de quelques-unes de celles-ci; le jaune et le bleu mêlés ensemble font du vert, le rouge et le bleu font du violet'. The results of mixing pairs of pigments selected from a wide range were described by Waller (1686). Waller calls yellows, reds and blues 'simple' colours, and greens and purples 'mean', but makes no generalization from his observations. There is no indication that he appreciated that one red, one yellow, and one blue, suitably chosen, would allow any colour to be reproduced.

The mixing of lights, as opposed to that of pigments, was first clearly described by Newton (1704), who published a diagram for predicting the results of colour mixture which, if it were exactly valid (but it clearly is not), would imply trichromacy. Newton recognized that his diagram, in the form in which he published it, was inexact, but he seems to have made no attempt to improve it. He did not point out its implications.

I cannot discover that any writer in the first half of the eighteenth century clearly appreciated the trichromacy of colour mixture, either for pigments or for lights. Certainly s'Gravezande, in his textbook of physics published in 1721, did not, for he wrote: 'Non omnium, qui in imagine solis oblonga observantur, colorum permixtio ad albedinem conflandam necessaria est. . . . Ex quatuor aut quinque colorum permixtione, justa servata proportione, albedo nascitur.' In fact, of course, two if suitably chosen (complementaries), or almost any three, suffice. However, in the second half of the century, without any author claiming to have made any new or original observation, trichromacy becomes universally accepted as a well-known fact. The earliest such acceptance that I can find is that of Lomonosov (1757). Lomonosov's statement of the results of mixing colours adds only a little to that of Mariotte, and it is evident that the distinction between mixtures of lights and of pigments is not appreciated: 'Why should Nature need special kinds of aether for orange, green, purple and other mixed colours, when she can construct orange from red and blue, and the other mixed colours from various other mixtures.' Nevertheless, it is clear from the context, and from the theory that he puts forward, that by 'various other mixtures' Lomonosov means mixtures of red, yellow and blue, which without white, black or any other supplement, suffice for making any colour.

   T. Mayer (1758) described the trichromacy of pigment mixture very exactly, treating it as a fact already well known, but without referring to any earlier writer. Mayer even proposed a quantitative trichromatic system for specifying colours, each colour being defined by the amounts of red, yellow and blue in it. Von Haller, in his large textbook of physiology of 1763, is equally lucid. There is a curious tendency of late writers, whose own statements are both clear and true, to attribute the discovery of trichromacy to much earlier writers whose published work does not well support the attribution. Thus Von Haller gives the credit, without evident reason, to s'Gravezande (1721), and Helmholtz (1863) to Waller (1686). Priestley (1772) puts the discovery of trichromacy, or something very like it, even earlier, for after referring to Hooke's *Micrographia* of 1665 he writes: 'Whereas before this period it had generally been supposed that there are three primary colours, Dr Hooke maintained that there are but two, blue and red, and that all the rest are composed of them.' In his account of Hooke's theory Priestley is accurate, but he can scarcely be correct in asserting that three primary colours were 'generally supposed' as early as this; neither Hooke himself, nor any other writer of the sixteenth, seventeenth or early eighteenth century known to me, gives any support for the opinion. It seems rather that the recognition of trichromacy grew gradually from the manifestly incomplete statements of Mariotte and Waller to the absolute clarity of Mayer and von Haller, and was not general until a decade or so before the date of Priestley's book.

Though Lomonosov wrote what is probably the earliest published statement of the empirical generalization that we call 'trichromacy', he does not claim to be its originator. Where he justly claims originality is in putting forward a theory to explain the facts of colour mixture. Lomonosov's theory is that there are three primary kinds of light, red, blue and yellow, and that all other kinds are mixtures of these three. There is a relic of alchemy in his further speculation that the three kinds of light are made of different kinds of particles in the aether, and that these have specific affinities for mercury, salt and sulphur. In passing, Lomonosov suggests that the sensitive particles of the retina also are of three kinds, and made of the same three materials. This theory explains satisfactorily all the facts of colour mixture for the normal human eye, but is very difficult to reconcile with the observations on refraction published more than fifty years earlier by Newton. Lomonosov does not discuss the difficulty; he ignores this aspect of Newton's work.

A theory similar in essentials to that of Lomonosov, but differing in emphasis, was published by G. Palmer in 1777. Where Lomonosov was mainly concerned with the nature of light, and only incidentally mentioned the eye, Palmer put equal stress on the threefold nature of light and of the sensitive structures of the retina. It remained for Thomas Young (1802a, b) to discard that part of Lomonosov's theory which conflicted with Newton's observations, and to suggest that there was a continuous series of kinds of light, but only three kinds of sensitive particle in the retina, each preferentially but not exclusively sensitive to light from one part of the spectrum. This suggestion has been the foundation on which all later theories of human colour vision have been built.

Many of these later theories are inferior in clarity and usefulness to Young's original suggestion. What Young wrote was simple and direct: 'Now, as it is almost impossible to conceive each sensitive point of the retina to contain an infinite number of particles, each capable of vibrating in perfect unison with every possible undulation, it becomes necessary to suppose the number limited; for

instance to the three principal colours, red, yellow and blue . . .
and that each of the particles is capable of being put into motion
less or more forcibly by undulations differing less or more from a
perfect unison . . . each sensitive filament of the nerve may consist
of three portions, one for each principal colour' (1802*a*). Later
(1802*b*) he added: 'In consequence of Dr Wollaston's correction of
the description of the prismatic spectrum . . . it becomes necessary
to modify the supposition that I advanced in the last Bakerian
lecture, respecting the . . . fibres of the retina; substituting red,
green and violet for red, yellow and blue.' Helmholtz, whose name
is commonly coupled with that of Young as originator of a theoretical explanation of trichromacy, substituted for Young's clear
and concrete 'sensitive particles of the retina' or 'sensitive fila-
ments of the nerve' the undefined theoretical concept of 'Grundempfindungen' (elementary sensations). Aubert (1865) severely
criticized Helmholtz for having in this way unnecessarily com-
plicated and confused the theory, and the criticism seems to me to
be justified, although in other ways, notably in his explanation
(1852) of the differences between additive and subtractive colour
mixture, Helmholtz made important contributions to our understanding of colour vision. There is no justification for calling the
hypothesis of three colour-sensitive mechanisms in the retina the
'Young–Helmholtz' theory. If it must be named eponymously it
should be the Lomonosov–Young theory, Lomonosov having been
the first to suggest it, and Young the first to rid it of that part
which, in its original form, was manifestly false, and leave it consistent with the evidence then available, and tenable, with minor
modification, up to the present day. We shall, however, not name
it after Lomonosov and Young, because what we wish to consider
is not Young's form of the theory, but several variants of it, each
as concrete and specific as their famous predecessor, but more
closely in accord with modern knowledge of the structure and
function of the retina.

## How exact is trichromacy?

*The fovea.* For fields of between 1° and 2° diameter, a number of
very thorough investigations of the colour-matching properties of
the normal eye have been made (König & Dieterici, 1886; Abney,
1913; Wright, 1929; Guild, 1931; Stiles, 1955; Stiles & Burch,
1959; Fridrikh, 1957). The earlier of these experimenters do not

record having detected any deviations from exact trichromacy, but Stiles (1955) notes that matches established on 2° foveal fields between sets of four lights differing widely in wavelengths are not always completely satisfactory, the match being slightly imperfect at the margin of the field when it is perfect at the centre and vice versa. This difficulty becomes more severe when larger fields are used, and with 10° fields it is very clear that, although the central part can be perfectly matched by using the three continuous controls of the trichromatic colorimeter if the outer part is ignored, or the outer part if the centre is ignored, no setting of the controls allows the centre and periphery to be matched at the same time. These differences between the matching properties of different parts of the foveal and near-foveal retina are largely, or perhaps wholly, due to screening of the receptors by macular pigment (see p. 140).

For fields of between 30' and 1° diameter it seems that trichromacy is exact. For smaller fields there are again departures from it, not, as with large fields, in the sense that three independent controls are insufficient for a satisfactory match, but that they are unnecessary; two suffice. This small-field dichromacy will be further considered on p. 240.

*Extrafoveal retina.* It is commonly supposed that the trichromacy found with foveal fields depends on cones only (see p. 257). If this supposition is correct, it is reasonable to expect that for extrafoveal fields, where rods also are illuminated, colour discrimination will, at least in favourable conditions, be tetrachromatic. Most of the work hitherto published has not supported this expectation. There is no indication in the descriptions of König (1896), von Kries (1896), Tschermak (1898), Wright (1946, p. 154), Gilbert (1950), Moreland & Cruz (1959) or Clarke (1963) that it is ever impossible to establish, for any one extrafoveal region of the retina, a perfect colour match between four lights, three of which can be varied continuously in luminance.

In all these experiments, the subjects compared two lights of different spectral composition which were presented simultaneously to neighbouring or nearly neighbouring areas of the retina. However, Bongard & Smirnov (1956) have found that if the two mixtures to be compared are presented successively to the same region of the retina, then no match can be made between mixtures of lights of wavelengths 435 nm, 490 nm, 592 nm and 630 nm; but addition of a fifth light allows a match to be made. Thus under these special conditions colour discrimination with extrafoveal fields is tetrachromatic, though with simultaneous presentation it is trichromatic, at least to a close approximation. I know of no published report of any attempt, successful or unsuccessful, to confirm this finding. Dr F. J. J. Clarke tells me that he knows of three independent unpublished attempts, all unsuccessful. Nevertheless, Dr Bongard has demonstrated the phenomenon to me (in Moscow in 1969), and I believe it to be real. At 160 cd m$^{-2}$ on a 1·5° field placed 5° from the fixation point the mismatch takes the form that on changing either from red + blue-green to yellow + violet or the reverse, there is a momentary apparent brightening. The steady appearances of the two fields do not, for me, differ.

## The three-channel hypothesis

Trichromacy can be explained if it is assumed that the pathway conveying information to consciousness from the photosensitive pigments of the retina is at any part of its course restricted to three channels in respect of colour sensitivity, the information carried by each channel at a given instant and for a given point on the retina being expressible as the value of one continuous variable. To state this three-channel hypothesis exactly in its most general form, it would be necessary to define very carefully the term 'channel'; but we are here mainly concerned with special forms of the hypothesis, in which the meaning of this term, or the more concrete ones that replace it, is unambiguous.

The three-channel hypothesis is not the only possible hypothesis consistent with trichromacy, but it is much the simplest. There will be no need to treat other hypotheses in detail, but a few of them will be briefly mentioned on pp. 209–210.

### Forms of the three-channel hypothesis

It is useful to distinguish three sub-hypotheses, which differ in their implications. The first, and nearest to the original suggestion of Thomas Young, is that there are only three photosensitive pigments in the region of retina concerned. It must be assumed either that each pigment takes part in only one kind of photochemical reaction capable of stimulating a receptor; or, if it takes part in more than one, either that the receptors do not qualitatively distinguish between the different photochemical reactions or that the action spectra for the different reactions of a given pigment are identical. It must also be assumed that all molecules of a given pigment are, to a close enough approximation, similarly situated in respect of filtration of the incident light by any coloured materials, or wavelength-dependent selection by any waveguide action that may be supposed to result from the structure of the receptors (see p. 252). Given these assumptions, which constitute what will be called the 'three-pigment hypothesis', trichromacy is already determined by the pigments, and it is not necessary (though it is simplest) to assume that there are only three kinds of receptor; if some receptors contain mixtures of pigments, there may be more than three kinds.

The second sub-hypothesis is that there are more than three pigments, but only three kinds of receptor, each receptor, even if it contains two or more different photosensitive pigments, transmitting a signal which for any one short time-interval can be wholly represented by the value of a single variable. The receptors thus determine a trichromacy although the pigments do not. This will be called the 'three-receptor hypothesis'.

The third sub-hypothesis, first clearly distinguished by Guild (1932), is that both the pigments and the receptors are of more than three kinds, but that in some more central part of the visual pathway, perhaps the retinal ganglion cells or the cells of the lateral geniculate body, there is a selective filter for information concerning colour, at which only three channels are available. For certain pairs of lights which differ in their effect on the receptors, the information that they do so is thrown away, and the message which is handed on to higher parts of the visual pathway fails to discriminate between them.

### Hering's theory

E. Hering (1875) put forward a theory which, as he later made clear in a long series of angry polemical papers against Helmholtz, Donders, von Kries and others, he regarded as sharply opposed to the three-channel hypothesis as expressed by Young and Helmholtz. The basic assumptions of Hering's theory (pp. 180–181 of his paper) are: 'Die sechs Grundempfindungen der Sehsubstanz ordnen sich zu drei Paaren: Schwarz und Weiss, Blau und Gelb, Grün und Rot. Jedem dieser drei Paare entspricht ein Dissimilirungs- und Assimilirungs-process besonderer Qualität, so dass also die Sehsubstanz in dreifach verschiedener Weise der chemischen Veränderung oder des Stoffwechsels fähig ist'. . . . 'Ich kann drei verschiedene Bestandteile der Sehsubstanz unterscheiden, welche ich als die schwarzweiss empfindende, die blaugelb empfindende und die rotgrün empfindende bezeichnen will'.

Even in the light of much collateral reading I find this a little obscure, as did Donders (1881) and other learned contemporaries. I think, however, that Hering was postulating three chemical substances in the retina, one (schwarzweiss empfindende) destroyed by any light and regenerated in darkness, the second (blaugelb empfindende) destroyed by yellow light and regenerated by blue, and the third (rotgrün empfindende) destroyed by red light and regenerated by green. Whether or not this correctly describes Hering's theory, it is easy to discover from his paper what he regarded as evidence in its favour. This is of three kinds: the subjective appearance of colours, successive contrast phenomena (Umstimmung), and colour blindness. The second and third of these can be shortly dismissed. In explaining successive contrast phenomena, Hering's theory is no better than Young's or Helmholtz's form of the three-channel theory; in explaining colour blindness, even as known in 1875, it is clearly worse. Hering's argument from the subjective appearance of colours, which he gives first and stresses most, is difficult to evaluate. The relevant fact is that to most people of European upbringing, red, yellow, green and blue seem to have a kind of simplicity that other colours do not, and that though other colours may be described as tinted with one of these four psychologically simple colours, or sometimes with two of them, they are never described as tinted with both red and green, or with both yellow and blue.

The arguments for Hering's theory may seem to some readers so feeble that they wonder why I mention the theory. The reason is that electrical recording has revealed among nerve cells of the visual pathway some that have properties roughly analogous to the 'Bestandteile der Sehsubstanz' that Hering postulated (see pp. 76, 84, 104, 119), and Hering is therefore regarded by some modern physiologists as having had prophetic insight. Now Hering's arguments from successive contrast phenomena and from colour blindness are bad, and it would be plainly improper to praise him for them. But his argument from the appearance of colours (a Class B phenomenon in the terminology of Chapter 5), though non-rigorous, is not necessarily bad; it is hard to judge because it comes from a kind of thinking that is outside the main tradition of natural science. My view is that we must postpone judgement until it has been determined whether those aspects of the subjective appearance of colours from which Hering argued are independent of the environment of the subjects during childhood. Among terrestrial natural objects, green and blue are commoner colours than blue-green, and many languages have simple words for green and blue and a compound word for blue-green. It might be that for a child brought up in an environment where all these three colours were equally common, and with a mother-tongue in which all three had equally simple names, the assertions from which Hering argued would not hold. If so, then the rough agreement (and it is only very rough) between Hering's theory and the discoveries of modern electrophysiology is a mere coincidence. But if the subjective simplicity of red, yellow, green and blue and the incompatibility of redness with greenness and of blueness with yellowness prove to be independent of upbringing, then Hering's argument is vindicated. The point concerns more than the reputation of a dead man; it will help us to decide whether in future to pay attention to arguments from Class B sensory phenomena, or to disregard them.

The trichromacy of colour mixture is ignored in all the writings of Hering that I have read, despite its apparent relevance to the controversies in which he took part. It is thus difficult to discover whether he understood that his theory is compatible with trichromacy, and can be adapted to provide an explanation of it. Hering's theory is, in fact, if I rightly interpret it, a special form of the three-pigment hypothesis, a form contradicted by modern biochemistry, but not absolutely incompatible with the results of any sensory experiment.

## Supplements to the three-channel hypothesis

A three-channel filter for information concerning colour, whether it lies in the pigments, in the receptors, or more centrally, suffices to explain trichromacy. The other parts of the visual pathway cannot contain fewer than three channels, for if they did, colour discrimination would be mono- or dichromatic. They may, however, contain more than three, not only peripherally to the filter, but also, as a result of interaction, centrally to it. Other possibilities for the central parts of the pathway are that there are three channels throughout, with the same spectral sensitivities everywhere, or three channels, but with different sets of spectral sensitivities in different parts. This last alternative is that postulated in 'zone' theories such as those of Donders (1881), Müller (1896), Judd (1949) and Hurvich & Jameson (1957). A 'zone' (Müller) or 'stage' (Judd) in the visual pathway is a structure or region to which information is transmitted and from which it is handed on to the next stage. A large number of zone theories can be constructed without coming into clear conflict with the results of any sensory experiments. Indeed, it is difficult to find even presumptive evidence from such experiments for preferring one zone theory to another, since the most peripheral three-channel stage already determines a trichromatic system of colour discrimination. The more central stages cannot add information which is not given to them, and cannot discard much without losing the trichromatic property. If a stage has merely reorganized the information received by it without adding or subtracting anything, it is not possible to infer anything about its properties from an examination of the same information after it has undergone further reorganization.

The discovery of opponent-colour properties in cells of the visual pathway (see pp. 76, 84, 104, 119) is sometimes held to have confirmed the truth of zone theories. But Donders, Judd and Hurvich clearly, and Müller if I rightly understand him, postulated just three colour-classes of cells in every zone. Electrophysiological findings do not clearly confirm this, and may be incompatible with it.

Inferences from the results of sensory experiments about parts of the visual pathway central to that at which the trichromacy of colour discrimination is determined are very insecure, and for this reason the discussion in the present chapter will be largely restricted to the properties and anatomical nature of the stage that determines trichromacy. On the three-channel hypothesis, this is the most peripheral stage that contains only three channels.

## Alternatives to the three-channel hypothesis

Schemes explaining trichromacy can be constructed in which no part of the visual pathway contains as few as three channels, because some information is thrown away at a stage in the pathway where other channels are carrying redundant information. There might, for example, be two different kinds of red-sensitive receptor, one of green-sensitive, and one of blue-sensitive. The two red-receptive channels converge on to a single channel at some more central stage of the visual pathway, perhaps the retinal ganglion cells; but here the blue and green channels have, by interacting, already ramified into more than two, so that there is nowhere a three-channel stage. The arbitrary complexity of this and similar schemes is evident, but they remain possible.

Another possibility is that there are four kinds of channel, but that their spectral sensitivities are related by a linear equation whose coefficients are independent of wavelength; or in general that there are $n+3$ kinds of channel with spectral sensitivities related by $n$ simultaneous linear equations. Stiles (1944) pointed out that the four mechanisms presumed to be present in extra-foveal retina might come sufficiently close to this condition to make colour discrimination trichromatic in all but a few especially unfavourable cases. The marked dependence of certain extrafoveal colour matches on field brightness (see p. 216 and Fig. 8.2) makes it clear that this cannot be a complete explanation of extrafoveal trichromacy.

### Testing the three-channel hypothesis

*The implications of its several forms*

The first form of the three-channel hypothesis, that of three photosensitive pigments, has, besides trichromacy, a number of other implications that can be tested. It implies that between multiples of any four lights a match can be established in which the two matching mixtures are alike in their effects on all structures in the retinal region concerned. Without making any further assumptions it can be proved, from the known impossibility of obtaining multiple discrete matches between the same components, that, except for the tolerances associated with the finite size of discrimination steps, all possible colour matches have this property (Brindley, 1957c). It follows that:

(1) non-identical lights that match (usually called 'metameric' lights) must have the same adapting effects, and in all circumstances give indistinguishable after-images;

(2) metameric lights must be able to replace one another as

constituents of other colour matches (Grassmann's third law; and

(3) lights that match in one state of adaptation must continue to match in all other states of adaptation, provided that the degree of adaptation is uniform over the region of retina which views both matching fields.

To the second and third of these deductions must be made the reservation 'unless the experimental conditions cause a change in the amount of screening pigment interposed between the test fields and some of the receptive pigment, or in the degree of selection by waveguide action of the light that reaches the receptive pigment'. This reservation has itself quantitative implications which can be tested. The first deduction holds without any reservation.

The three-receptor form of the three-channel hypothesis, that is that there are more than three pigments, but only three kinds of receptor, shares with the three-pigment form the implication that metameric lights must be able to replace one another as constituents of other colour mixtures, provided that in a receptor containing two pigments the effects of photochemical reactions in the two pigments add in a linear manner. Such direct additivity of the effects of two photochemical reactions in the same receptor is likely on physiological grounds, but not absolutely necessary.

On the three-receptor hypothesis, the deduction that metameric lights should have the same adapting effects is valid only if several further assumptions are made. The hypothesis thus remains tenable whether the property is true or false. For the stability of colour matches to adaptation, however, the three-receptor and three-pigment hypotheses differ sharply in their implications. Ophthalmoscopic densitometry proves (see p. 40) that adapting lights, if bright, can bleach away a large fraction of the cone pigments of the retina. If a receptor contains two pigments whose action spectra for bleaching are different, an adapting light that preferentially bleaches one pigment will alter the spectral sensitivity curve of the receptor in one direction, and one that preferentially bleaches the other pigment will alter it in the opposite direction (see Fig. 8.1). If the trichromacy of colour matching is determined by receptors and not by pigments, both of these adapting lights must disturb colour matches, and in opposite directions.

The third form of the three-channel hypothesis, namely that

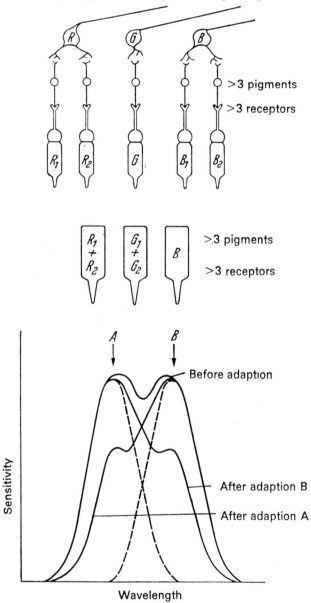

FIG. 8.1. An implication of the hypothesis that there are more than three foveal receptor pigments. Whether these are in separate receptors (*top sketch*) or mixed in three kinds of receptor (*middle sketch*), colour matches will be disturbed by adaptation, and in opposite directions for different adapting colours (from Brindley, 1959*b*).

there are more than three kinds both of pigments and of receptors, converging on to three classes of neurons in respect of spectral sensitivity at some more central stage in the visual pathway, shares none of the special implications of the three-pigment hypothesis. Metameric lights do not in general affect the receptors alike, so there is no necessary reason why they should have the same adapting effects or after-images; though it is possible to devise *ad hoc* supplementary hypotheses predicting that they will. Metameric lights need not necessarily be able to replace each other as constituents of other colour matches. They will be able to do so only if the response of each channel at the first three-channel stage in the visual pathway is determined by a linear function of the rates of photolysis of all kinds of receptive pigment connected to it. It would require a mechanism very different from any known neural mechanism to achieve this with accuracy over a wide range of intensities of stimulus. There is no reason to expect colour matches to be stable to adaptation; on the contrary, there is the same reason as with the three-receptor hypothesis for expecting that adapting lights of different spectral composition will disturb colour matches in different ways.

*Special form of the three-pigment hypothesis where the density of pigment is high*

In the original statement of the three-pigment hypothesis, it was pointed out that, provided that all molecules of a given pigment are, to a close enough approximation, similarly situated in respect of filtration of the incident light, the presence of three receptive pigments suffices to determine a trichromacy, even if there are more than three kinds of receptors, some of them containing mixtures of pigments.

If the density of receptive pigment in the receptors is high, then the molecules of a given pigment cannot all be similarly situated in respect of filtration of incident light, and two lights which have identical effects on all three pigments when these are exposed in thin layers at low concentration, may not have identical effects on the receptors in the eye. In these circumstances, three pigments are insufficient to determine a trichromacy: it is necessary that there be three kinds of receptors as well, or at least that any pigment which occurs in high density be strictly segregated, either alone or with a fixed proportion of other pigment or pigments, in one kind of receptor. This modified three-pigment hypothesis shares the implications of the simple three-pigment hypothesis concerning Grassmann's third law, the identical adapting effects of metameric lights, and the stability of colour matches to degrees of adaptation insufficient to bleach a substantial fraction of any receptive pigment; but concerning the effect on colour matches of adaptation to lights bright enough to bleach the pigments, it has implications of its own, which will be considered on p. 221.

### Experimental evidence

*The after-images and adaptive effects of metameric light.* For the fovea, the available experimental evidence is not very complete, but probably indicates that the after-images are always indistinguishable. Cohen (1946), in an experiment described very briefly, found no difference between the adaptive effects of a monochromatic yellow and those of a red-green mixture which matched it. Hartridge (1948), adapting to light from the sky which had passed through filters transmitting either yellow only or a mixture of red and green, found significant differences. I repeated Hartridge's experiments under the same conditions (Brindley, 1960*b*), and could confirm his observations in almost every detail, but only when the green component of the red-green mixture contained significant amounts of light of wavelengths less than 540 nm, so that the mixture was obviously less saturated than the pure-yellow. When the red-green mixture was made to contain no light of wavelength less than 540 nm, so that the difference in saturation was extremely small, no difference in adaptive effects could be found. It seems likely, therefore, that Hartridge's result was due to failure to match the saturations, and is irrelevant to the present question. In a short series of experiments done with a divided circular foveal field of 1° diameter (Brindley, 1960*b*), no difference between the after-images of metameric lights could ever be detected, whether they were viewed against a dark background or against white or monochromatic backgrounds of various luminances.

For extrafoveal retina, it seems fairly certain that a careful search would reveal metameric pairs with differing adaptive effects, but I can find no published account of such a finding.

*Can metameric lights be substituted for one another as constituents of other colour matches?* Grassmann (1853) was the first to assert that such substitutions could be made without disturbing the matches. His third law of colour mixture stated 'dass . . . zwei Farben, deren jede constanten Farbenton, constante Farbenintensität, und constante Intensität des beigemischten Weiss hat, auch constante Farbenmischung geben, gleichviel aus welchen homogenen Farben jene zusammengesetzt seien'. Experimental evidence is available both for foveal and for extrafoveal fields, much of it relating to the special case of the stability of colour matches to the brightening or dimming of all constituents by a constant factor. For foveal fields of 1° 20' diameter or less, von Kries (1896), Tschermak (1898) and Trezona (1954) agree in finding no breakdown of colour matches on proportional dimming. Stiles (1955) in a single experiment on a 2° field finds a very small breakdown not certainly outside the very narrow limits of his experimental error (see Fig. 8.2). More general tests of the substitution property,

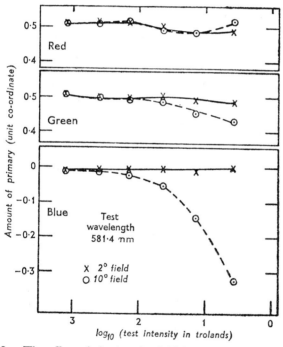

FIG. 8.2. The effect of altering the field intensity on the proportions of red, green and blue primaries (649, 526 and 445 nm) required to match a fixed spectral yellow (from Stiles, 1955).

carried out by Blottiau (1947) on a 2° field and by Trezona (1953, 1954) on a 1° 20′ field, have shown deviations, but these are nearly always small compared with a just discriminable colour difference, and it is very doubtful whether they can be taken to exclude a hypothesis implying that substitutions can be made. At very high luminances, in view of the known disturbance of colour matches by adaptation (see below) significant deviations from Grassmann's third law must exist, though they have not yet been directly reported.

For extrafoveal fields, large deviations from Grassman's third law are established beyond doubt. They were first demonstrated by König (1896) and von Kries (1896), whose findings have been many times confirmed and amplified (for references see Lozano & Palmer, 1967, and Clarke, 1963). They occur over a wide range of luminances, and are greatest at luminances where other evidence suggests that rods as well as cones are contributing to extrafoveal vision.

*The stability of colour matches to adaptation.* Conspicuous changes in the appearance of colours can be produced by adaptation to coloured lights without upsetting colour matches made between lights of different spectral composition; indeed von Kries (1878), having deduced from Helmholtz's form (1867) of the three-channel theory that no colour match should be disturbed by adaptation, investigated, as a test of the theory, matches of ordinary white light against mixtures of pairs of complementary spectral lights before and after adaptation to various coloured lights, and failed to detect any alteration in the amounts or wavelengths of the complementaries required. However, with very bright adapting lights, as Wright (1936) showed, colour matches can be disturbed. It can easily be proved that this disturbance is not due to production or destruction of a general screening pigment lying in front of all the receptors equally, for the factor by which a given adaptation alters the ratio of amounts of fixed red and green primaries required to match a variable test wavelength differs for different test wavelengths (Brindley 1953; see Fig. 8.3). A general screening pigment could alter this ratio only by absorbing light at the wavelengths of the red and green primaries, and such absorption would necessarily be independent of test wavelength.

A disturbance of colour matches not due to production or

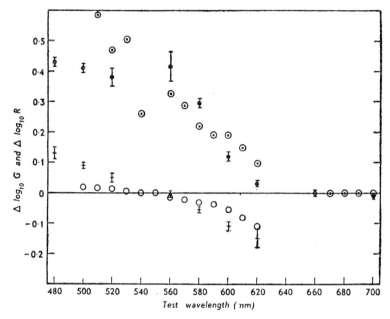

Fig. 8.3. Changes produced by adaptation to yellow light (equivalent to 578 nm) at 1300 mL for 30 sec in the amounts of red
(680 nm; ↕) and green (550 nm; ↕) primaries required to match
test lights of various wavelengths. ⊙ and ○ show the values
calculated on the hypothesis that the changes are due to bleaching
of the red-receptive pigment, this being initially present in very
high density (from Brindley, 1953).

destruction of a general screening pigment is inconsistent with the
simple form of the three-pigment hypothesis. Both the hypotheses
postulating more than three pigments predict, in their simple
forms, that if an adapting light of one wavelength disturbs a colour
match in one direction, it should be possible to find an adapting
light of some other wavelength which disturbs it in the opposite
direction. This, however, is not possible. Figure 8.4 shows the
disturbing effects of a number of adapting lights on one colour
match. All the adapting lights, whatever their wavelength or
brightness, cause an increase in the amount of red required in the
colour match and a decrease in the amount of green, the increase in
red being always about five times the decrease in green. The
simple forms of all the hypotheses under consideration are thus

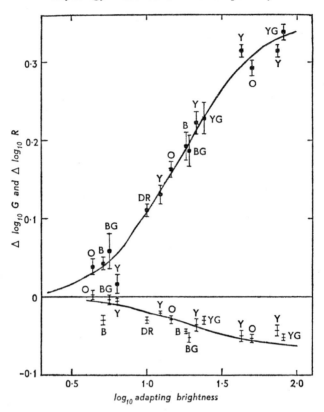

FIG. 8.4. Changes produced by adaptation to various lights in the amounts of red (680 nm; $\text{Ⅰ}$) and green (550; $\text{Ⅰ}$) primaries required to match yellow light of wavelength 580 nm. The wavebands of adapting lights were isolated by means of filters, and were equivalent to 480 nm (B), 497 nm (BG), 550 nm (YG), 606 nm (O), and 680 nm (DR) (from Brindley, 1953).

contradicted, and we must examine the modifications that they need if they are to explain the facts.

The hypothesis of more than three pigments and more than three classes of receptor can be reconciled with the similar disturbing effects on colour matches of adaptation to lights of different wavelengths only if, wherever two classes of receptor converge on to a single class of neuron at some more central stage in the visual pathway, one class is so much more adaptable than the other that it alone is substantially adapted, even when the adapting wavelength is such that the other class is the more sensitive; and in order to explain the relation between the intensity of the adapting light and the degree of disturbance of colour matches (Fig. 8.5) even the

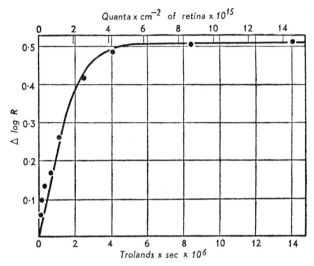

FIG. 8.5. Changes produced by adaptation to yellow light at various intensities for 10 sec in the amounts of red primary required (together with a fixed amount of green) to match a fixed spectral yellow (570 nm) (from Brindley, 1955b).

more adaptable class of receptors must be unaffected in sensitivity by amounts of light less than about $2 \times 10^5$ trolands × seconds. Each of these conditions is relatively unlikely, and together they add up to a high degree of implausibility. Further evidence against an explanation of trichromacy in terms of more than three classes of receptors is that a given quantity of light has the same effect in disturbing a colour match, whether it is concentrated within one second or spread out over ten seconds (see Fig. 8.6). This is to be expected if the disturbance depends directly on a photochemical reaction, but would be surprising if it depended on an adaptive process in receptors or neurons.

To reconcile the three-receptor hypothesis (i.e. that of more than three pigments contained in three classes of receptor) with the similar disturbing effects on colour matches of different wavelengths

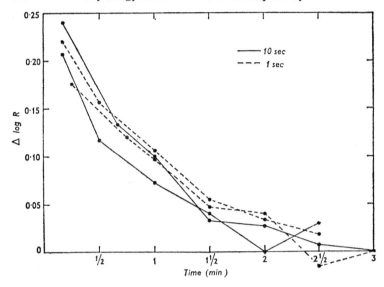

FIG. 8.6. Time-course of recovery of a colour match disturbed by adaptation to $8 \cdot 5 \times 10^5$ trolands $\times$ seconds concentrated within 1 sec and spread over 10 sec (from Brindley, 1955b).

of adapting light, it is necessary to suppose that, wherever two pigments are contained in the same receptor, one pigment is so much more photolabile than the other that it alone is substantially bleached, even when the adapting wavelength is such that the other pigment contributes more to excitation of the receptor. This is not as unlikely as the corresponding condition for the hypothesis of more than three classes of receptor. The less bleached pigment could resemble a sensitizer in a photographic emulsion, handing on the energy of each absorbed quantum immediately to some other molecule and returning to its original state, so that it would not, by ordinary criteria, be photolabile.

The three-pigment hypothesis allows colour matches to be disturbed by adaptation only if the adaptation produces or destroys a screening pigment in front of the receptors. We have seen that no general screening pigment, lying in front of all receptors equally, can explain the results. In reptiles and birds, coloured oil droplets are present in the cones, each droplet appearing to act as an individual colour filter for the outer segment of the cone in which it lies; but such structures have never been observed in mammals, and there is strong evidence from sensory experiments against their existence in man (see p. 250). There is, however, an individual screening pigment which is undoubtedly present in each mammalian cone: the receptive pigment itself. If this is photolabile and is present in sufficient density for the pigment in the posterior part of each outer segment to be significantly screened by that in the anterior part, it must necessarily produce effects analogous to those observed. The implications of such 'self-screening' of a photolabile receptive pigment have been examined in detail (Brindley, 1953, 1955b), on the basis of two simplifying assumptions: that the red-receptive and green-receptive pigments are entirely segregated in different receptors, and that self-screening, or at least alteration of self-screening by light-adaptation, is appreciable only for one of them. Given these assumptions, it must be the red-sensitive pigment that has the high density, and its density must be 0·98 for the one subject investigated in detail; required densities for other normal subjects always exceed 0·5. Of the two assumptions, the first must be at least nearly true if three pigments are to imply a trichromacy, and the observations of Marks, Dobelle & MacNichol (1964) and Brown & Wald (1964) on single human cones support it. I argued (1953) in favour of the second assumption from the close similarity of the effects of adapting lights of different wavelengths on a colour match (see Fig. 8.4). But this part of the argument is much less good than I supposed; the observations of Fig. 8.4 do not sharply conflict with the hypothesis, tentatively suggested and well discussed by Enoch & Stiles (1961), that both the red-sensitive and the green-sensitive pigment have densities between 0·5 and 1.

Dunn (1969) quotes from P. Liebman, and verifies for the rods of the gecko, the generalization that the optical density along a rod or cone outer segment is 0·015 per micron. The empirical basis for such a generalization is slighter for cones than for rods, but its application to cones fits the electron-microscopical structure well. If it is valid, it predicts a density of 0·75 for human foveal cones, but only about 0·3 for human extrafoveal cones. The obvious implication, on the self-screening hypothesis, that the disturbance of colour matches by adaptation should be much less for extrafoveal fields, remains to be tested.

Grounds for doubting arguments from spectrophotometry, according to which the densities in cones are very much too low to explain Figs. 8.3–8.6 and 8.19 in terms of self-screening, are mentioned on p. 44.

*Objective measurements of human cone pigments.* The observations of Rushton by ophthalmoscopic densitometry, and of Brown & Wald and Marks, Dobelle & MacNichol on single human cones (see pp. 40–46) show that there are at least three such pigments, that their difference spectra and the action spectra for bleaching them agree at least roughly with expectations from sensory experiments, and that they are at least mainly segregated in different classes of cones, erythrolabe in red-sensitive cones, chlorolabe in green-sensitive, and cyanolabe in blue-sensitive. The observations do not prove that there are only three cone pigments; if there are more, the ones undiscovered by objective observations must be either scarce or not manifestly photolabile, but neither scarcity nor lack of manifest photolability would exclude their being physiologically important. Nor do the objective observations prove that the segregation of each cone pigment in a special class of cones is absolute, though absolute segregation would almost certainly be the most economical way of using them for colour discrimination, and we can expect evolution to have achieved efficiency.

*Results of recording from single neurons of the visual pathway.* The observed properties of single cells (see pp. 76, 83, 104, 119) do not yet help us to distinguish between the three forms of the three-channel hypothesis, or between that hypothesis and the alternatives to it. If more fully investigated they might so help, but I suspect that such an investigation would be very laborious and only slightly rewarding.

*Summary of evidence.* We can draw the following conclusions.
The evidence from the after-images of metameric lights is difficult to reconcile with a trichromacy determined centrally to the receptors. It does not distinguish between the three-receptor and three-pigment hypotheses.

For the fovea, the validity of Grassman's third law is very difficult to reconcile with trichromacy determined centrally to the receptors, and puts constraints on any hypothesis that postulates more than three receptive pigments. For extrafoveal retina, the invalidity of Grassman's third law excludes the three-pigment hypothesis, and strongly suggests that there the trichromacy *is* determined centrally to the receptors.

The lack of disturbance of colour matches by moderate chromatic adaptation further strengthens the case against the determination of foveal trichromacy centrally to the receptors.

The nature of the disturbance of colour matches by extreme chromatic adaptation proves that if there are more than three receptive pigments, all but three of them must be practically photostable. If the explanation of this disturbance in terms of self-screening is correct, there must be both three pigments and three kinds of receptor.

Objective spectrophotometric measurements support the view that there are both three cone pigments and three kinds of cone, but are not full enough or accurate enough to establish it beyond doubt.

## The spectral sensitivities of the three channels

Whether the three channels determining trichromacy are pigments, or receptors, or structures that lie at some more central level in the visual pathway, the colour-matching properties of the eye provide some information about their spectral sensitivities. It has been generally accepted since König & Dieterici (1886) that the reciprocals of the spectral sensitivities of the channels determining trichromacy must be linear functions (with coefficients independent of wavelength) of the amounts of three specified primaries required to match constant energies of light at the wavelengths for which the spectral sensitivities are to be determined. The basis of this assumption is the well-known linear property of transformations from one set of real primaries to another (see, for example, Wright, 1946, Chapter VIII), which can easily be deduced from Grassman's third law. To apply the linear property to the spectral sensitivities of the three channels determining trichromacy, we must postulate unreal or 'extra-spectral' primaries, each consisting of a positive amount of one spectral light added to negative amounts of one or two others, and argue that there must be three primaries, real or unreal, that obey the linear transformation rule and are such that each stimulates one only of the three channels determining trichromacy. These are the 'fundamental stimuli' of Wright (1934) and other authors. If we know what they are, we can calculate the amounts of them required to match equal energies of light throughout the spectrum, and the reciprocals of these amounts must be the spectral sensitivities of the three channels. The colour-mixing properties of the eye do not suffice to determine what the fundamental stimuli are, and all of the many attempts that have been made to estimate them (for references see Wright, 1946, Chapter XXX) involve assumptions that may be false. The various estimates are not very closely similar, but they have some features in common. In particular, nearly all of them agree in placing the red and green fundamental stimuli on the straight line to which the long-wavelength end of the spectral locus in the colour triangle closely approximates (see Fig. 8.7). They consequently agree precisely (when they are based on the same colour-matching data) in their estimates of the spectral sensitivity of the blue-sensitive channel, for in linear transformation to any set of primaries the spectral distribution of each primary is uniquely determined by the position in the colour tri-

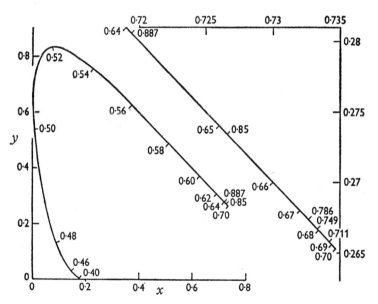

FIG. 8.7. The locus of the spectrum in the C.I.E. chromaticity diagram, showing the return of the locus along itself at very long wavelengths. On the right is shown a twenty times magnification of this part of the locus (from Brindley, 1955c).

angle of the line joining the other two; it depends neither upon its own position in the triangle nor upon where on their straight line the other two lie.

Figure 8.8 shows the spectral sensitivities of the three channels as estimated by Pitt (1944). Pitt derived the positions of the red and blue fundamental stimuli from the assumptions that in the forms

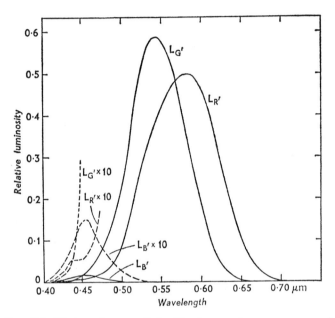

FIG. 8.8. Pitt's estimates of the 'fundamental response curves', i.e. the spectral sensitivities of the three mechanisms responsible for the trichromacy of colour matching (from Pitt, 1944).

of colour-blindness known as protanopia and tritanopia (see p. 227), the red and blue channels respectively are missing, and those channels which remain are normal in their spectral sensitivity. These assumptions are very probably correct (see p. 228). For the green fundamental stimulus, Pitt used the estimate of Walters (1942; see p. 238), derived from the effects of adaptation upon apparent colour.

The sceptical reader may reasonably ask how securely established is the generally accepted rule that the reciprocals of the spectral sensitivities of the three channels must be linear functions of the amounts of three primaries required to match equal energies of spectral lights. This rule was introduced above through the rather difficult concept of unreal 'fundamental stimuli', because they provide one useful way of looking at the problem, and are firmly established in its literature. The rule is not directly demonstrable from experimental observations, and an assumption must be made in order to prove it. The assumption can be expressed in the form of an extrapolation of Grassman's third law beyond its verifiable instances, but the argument from this extrapolation is very abstract. A more concrete approach is from the way in which the channels determining trichromacy respond to mixtures of lights. It is sufficient to assume that if any lights $A$ and $B$, of different spectral composition, have the same effect on a given channel, then $(A+C)$ has the same effect as $(B+C)$, where $C$ is any other light; and conversely that if $(A+C)$ has the same effect as $(B+C)$, then $A$ has the same effect as $B$. A proof based on this assumption is given in Brindley, 1960$a$.

For a trichromacy determined by pigments, the property assumed above has good theoretical as well as experimental foundation, and the generally accepted rule that the reciprocals of the spectral sensitivities are linearly related to the colour-matching functions is likely to be exactly valid. If foveal trichromacy is not determined by pigments, it may be only an approximation, though probably a close one. Where, as in extra-foveal retina, Grassman's third law does not hold, there is no reason for supposing that it is even approximately true.

### Colour blindness

Abnormalities of colour vision of genetic origin, generally known as colour blindness, are common in men, but much less common in women (roughly 8% and 0·4%). The abundant types, namely protanopia, deuteranopia and the red-green forms of anomalous trichromacy, are inherited as totally sex-linked recessive characters which are either allelomorphic or closely linked to each other.

Colour-blind subjects are customarily divided into a small number of classes, the boundaries between which are fairly sharp. The majority of subjects fit clearly into this classification, but there are a few, including the case described by Richardson-Robinson (1923), to whom it seems to be inapplicable. In the brief account of the recognized classes of colour-blind subjects which follows, recent and convenient references are given, without necessarily respecting priority of discovery.

*Monochromats* are defined as those who are quite unable to distinguish colours; any colour can be matched by them with any other, if the intensities are suitably adjusted. A well-known and apparently circumscribed class is constituted by the so-called rod monochromats (Falls, Wolter & Alpern, 1965) whose visual acuity is poor, and whose spectral sensitivity, in both dark-adapted and light-adapted states, corresponds to that of rhodopsin. Despite this, their retinas contain cones. Presumably these cones are either filled with rhodopsin or functionless. The shape of the rod monochromats' dark-adaptation curve favours the former alternative.

$\pi_1$ *cone monochromats* (Alpern, Lee & Spivey, 1965) also have poor visual acuity. Their dark-adaptation and their spectral sensitivity in the dark-adapted state are normal, but their foveal spectral sensitivity in the light-adapted state is that of the blue-receptive cone system ($\pi_1$; see p. 234).

$\pi_4$ *cone monochromats* (Weale, 1953b, who calls them simply 'cone monochromats', since the $\pi_1$ type was not then known) have normal visual acuity, and their foveal spectral sensitivity in the light-adapted state has its maximum at about 545 nm, like that of protanopes. Their dark-adaptation is probably normal, and their extrafoveal spectral sensitivity in the dark-adapted state is that of rhodopsin (Baumgardt, 1955).

Between 1 in $10^4$ and 1 in $10^5$ of both men and women are rod monochromats. Both forms of cone monochromacy are even rarer.

*Dichromats* can match all colours with suitable mixtures of two primaries. They are customarily divided into protanopes, deuteranopes and tritanopes. The protanopes and deuteranopes (Wright, 1946, Chapter XXVI) have very little colour discrimination at the red end of the spectrum, and the spectral colour that they cannot distinguish from white (neutral point) is in the blue-green, about 495 nm for protanopes and 500 nm for deuteranopes. The primaries that they require to match the whole spectrum by positive mixture are red, orange or yellow (which are equivalent, because indistinguishable) and violet of wavelength 430 nm or less. The spectral sensitivity of deuteranopes is almost normal. Protanopes, on the other hand, are abnormally insensitive to the red end of the spectrum. About 1% of men are protanopic, and about 1% deuteranopic. Tritanopia (Wright, 1952) is a much less common condition, affecting probably between 1 in $10^4$ and 1 in $10^5$ of the population. It is no rarer, or not much rarer, in women than in men, and the published pedigrees are compatible with autosomal dominant inheritance (Henry, Cole and Nathan, 1964). Tritanopes have good colour discrimination at the red end of the spectrum, but very little ability to distinguish blue from green. Their neutral point is in the yellow, about 575 nm. To match the whole spectrum by positive mixture, they require red and blue (460 nm) primaries, not red and violet, and the extreme violet is to them almost (perhaps quite) a second neutral zone.

Protanopia, deuteranopia and tritanopia are to a close approximation reductions of normal trichromatic vision; that is, though the colour matches that dichromats accept are often very far from being matches for the normal eye, matches that are accepted by normal subjects are also accepted, or very nearly accepted, by dichromats. From this it has been generally supposed since the end of the last century that protanopia and tritanopia are due to lack of the red-sensitive and blue-sensitive respectively of the three channels that in normal subjects determine trichromacy. The corresponding hypothesis for deuteranopia, i.e. that it is due to lack of the green-sensitive channel, was also proposed in the last century, but was less generally accepted, because of a real but not insuperable difficulty in reconciling it with the spectral sensitivity of deuteranopes. Rushton (1963a, 1965a) has shown that the red-sensitive and green-sensitive foveal pigments erythrolabe and chlorolabe are absent or unmeasurably scarce in protanopes and deuteranopes respectively, and has thereby made it extremely probable that protanopia and deuteranopia are indeed each due to loss of one of the normal three channels. That tritanopes similarly lack the blue-sensitive pigment cyanolabe remains likely, but there is no evidence for or against. It is a natural further speculation that $\pi_1$ cone monochromats lack erythrolabe and chlorolabe, and $\pi_4$ cone monochromats lack erythrolabe and cyanolabe. For the $\pi_4$ cone monochromats, evidence against such a hypothesis has been given by Weale (1959) and Gibson (1962), but does not absolutely convince

me. Ikeda & Ripps (1966) found the electroretinographic spectral
sensitivity curve of a $\pi_4$ cone monochromat to agree with his
spectral sensitivity curve, as would be expected if erythrolabe
and cyanolabe were lacking, or at least ineffective in exciting
cones.

*Anomalous trichromats* (Wright, 1946, Chapter XXV) make up
nearly 6% of the male population. They resemble normal subjects
in that they require three primaries to match all colours by
mixture; but they require them in abnormal proportions, so that
they neither accept normal colour matches nor make matches
which normal subjects accept. In the forms of anomalous trichromacy that have been extensively investigated, the abnormality of
colour matches affects mainly the red end of the spectrum, and
there is usually also a defect of discrimination in this part of the
spectrum. The degree of disturbance of the mean setting of a
colour match (the 'Rayleigh match' of yellow against a mixture of
red and green is usually taken as representative) and the scatter of
readings (which gives a measure of the defect of discrimination)
both vary greatly from one anomalous trichromat to another, and
are not closely correlated; a severe defect of discrimination is
sometimes found with only a small disturbance of the mean match.
Anomalous trichromats thus form a heterogeneous group, for the
adequate subdivision of which at least two independent measurements are required. The customary division into protanomalous
and deuteranomalous, according to whether an excess of red or
green is required in the Rayleigh match, does not always correctly
predict the shape of the spectral sensitivity curve, but most protanomalous subjects, as diagnosed on the Rayleigh match, are
abnormally insensitive to red, and most subjects who are deuteranomalous by the same criterion have normal spectral sensitivities.

Besides the well-known red-green forms of anomalous trichromacy, there are forms ('tritanomaly') in which matching and/or
discrimination are more abnormal at the blue end of the spectrum
than at the red. Very little quantitative information is available
about these. Such as there is has been collected by Engelking
(1926) and Jaeger (1955).

*Acquired colour blindness.* Though non-progressive colour
blindness of genetic origin is very much commoner, defects of

colour vision can also be acquired in post-natal life as a result of disease of the retina, optic nerve or visual cortex. The literature of these acquired defects has been well reviewed by François & Verriest (1957), and further cases are reported by Grützner (1961). It has never been found that colour matches formerly accepted cease to hold, except in conditions where this could readily be attributed to pigmentation of the transparent media of the eye. The defects not so attributable have usually been accompanied by severe disorders of visual acuity or visual field, and have consisted of a general increase in the size of discrimination-steps, often affecting the blue end of the spectrum more than the red. Occasionally the disorder amounts to full tritanopia, but this differs from congenital tritanopia in that discriminations of colour along the red-green axis and of brightness are also impaired.

## Discrimination steps and incremental thresholds

In the arguments already considered, it has been assumed that ideally, for any four lights one of which is fixed, there is only one set of quantities of the other three from which a colour match can be made; and the fact that, in practice, the quantities can be varied over small ranges without detectably disturbing the match, has been regarded only as a source of experimental inaccuracy in determining parameters which ideally have exact values. From this point of view we saw that, since trichromacy is already determined by the most peripheral three-channel stage in the visual pathway, little or nothing can be learnt from sensory experiments about more central stages, for these cannot add anything to the information handed on to them, and if they reorganize it, do nothing thereby which could not in principle be reversed at some still later stage. There is, however, one thing that stages in the visual pathway central to that determining trichromacy might do that could be detected by sensory experiments: they might throw away small amounts of information, thereby becoming the determining factor in the size of discrimination steps.

It is distasteful to a biologist to postulate a biological transmission line which arbitrarily fails to transmit part of the information given to it; but the eye is not solely an organ for the discrimination of colour, and it is not unreasonable that the visual pathway should sometimes discard information about colour in order to reserve more channels for transmitting information of other kinds. The pathway from the extrafoveal part of the retina to the brain certainly discards much of the information about spatial pattern available in the responses of its receptors; and if a comparable filtering of colour information occurs, discrimination steps may not be determined wholly by the structures that determine trichromacy, but also by more central parts of the visual pathway.

Extensive and accurate data on the size of discrimination steps have been published, notably by MacAdam & Brown (MacAdam, 1942, 1949b; Brown & MacAdam, 1949; Brown, 1957), and by Stiles (1939, 1949, 1953, 1959; Boynton, Ikeda & Stiles, 1964). MacAdam and Brown measured the errors made in attempting to establish a trichromatic match. In their early experiments, the brightness was held constant, but in the later experiments, discrimination steps are measured in all directions in brightness-colour space from the initial chromaticity and brightness. The locus of a point which lies one discrimination step from a fixed point in brightness-colour space is found to be, within the limits of accuracy of the determination, an ellipsoid.

Stiles's measurements take the form of determining the least amount of a given spectral light which must be added to a background of another spectral light (of the same or different wavelength) for the increment to be detected. The brightness of the background is varied over a very wide range. The kind of pattern into which the results fall is shown in Figs. 8.9 and 8.10. Each curve shows, for a given background wavelength ($\mu$) and test wavelength ($\lambda$), the effect on the incremental threshold of varying the brightness of the background. Many of the curves where $\lambda \neq \mu$ are divisible into two sections, each section having the same shape as a part of the simple curve obtained when $\lambda = \mu$, but displaced in relation to the axes without rotation. Varying the background wavelength (Fig. 8.9) is found to cause displacement of the curve parallel to the axis of log (background intensity), and varying the test wavelength (Fig. 8.10) to cause displacement parallel to the axis of log (test intensity). The law relating degree of displacement to the wavelength is different for the different sections of the curves; but for a given section, as far as can be ascertained, the same law relates vertical displacement to test wavelength and horizontal displacement to background wavelength. It thus defines a single characteristic spectral sensitivity for the section.

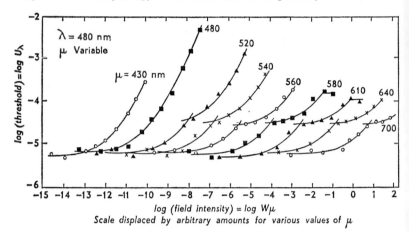

FIG. 8.9. Effect on the curve of log (threshold) against log (background intensity) of changing the wavelength of the background ($\mu$), and keeping the test wavelength ($\lambda$) constant. Foveal test stimulus of 1° diameter (from Stiles, 1949).

On the basis of these measurements of increment thresholds Stiles (1946) has put forward a general theory of brightness-colour discrimination of great formal simplicity, which not only provides a complete interpretation of the main features of his own measurements, but also agrees well with published data on the size of hue-discrimination steps. The theory, in its original form, postulates three mechanisms, the incremental threshold of each of which depends only on the extent to which it is stimulated by the background, and is entirely independent of the activity of the other two mechanisms. The threshold amount of a given incremental stimulus for the eye as a whole is given by

$$\left(\frac{1}{U_\lambda}\right)^2 = \left(\frac{1}{U_{\lambda_r}}\right)^2 + \left(\frac{1}{U_{\lambda_g}}\right)^2 + \left(\frac{1}{U_{\lambda_b}}\right)^2$$

where $U_{\lambda_r}$, $U_{\lambda_g}$ and $U_{\lambda_b}$ are the threshold amounts of the given stimulus for the three hypothetical mechanisms separately, and $U_\lambda$ that for the eye as a whole. Thus unless two of $U_{\lambda_r}$, $U_{\lambda_g}$ and $U_{\lambda_b}$ are nearly the same, $U_\lambda$ is approximately equal to the smallest of them. In Stiles's experiments (up to 1959), $U_{\lambda_r}$, $U_{\lambda_g}$ and $U_{\lambda_b}$ are all of the same sign: the stimulus, when just large enough to be

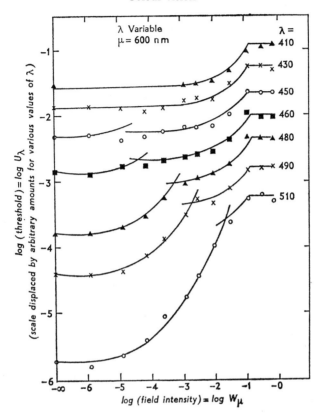

FIG. 8.10. Effect on the curve of log (threshold) against log (background intensity) of changing the wavelength of the test stimulus ($\lambda$), and keeping the background wavelength ($\mu$) constant. Foveal test stimulus of 1° diameter (from Stiles, 1949).

detected by a given mechanism, will always constitute an incremental threshold for it. In more general discrimination experiments (Boynton, Ikeda & Stiles, 1964) the stimulus may approach an incremental threshold for one mechanism but a decremental for another. The original theory assumed, in accord with the second-degree equation which expresses it, that summation between two mechanisms towards the attaining of threshold takes no account of sign, i.e. of whether the stimulus represents an increase in the degree of stimulation of both mechanisms, or an increase in one and a decrease in the other. For side-by-side comparison, this assumption is supported by the ellipsoidal form of confusion regions in colour space found by Brown and MacAdam. But for successive comparison, the experiments of Boynton, Ikeda & Stiles (1964) clearly show it to be false. An incremental stimulus to one mechanism and a decremental stimulus to another summate, i.e. both together are seen when each separately is of just subliminal strength; but incremental stimuli to two mechanisms, or decremental stimuli to two mechanisms, generally reveal mutual inhibition, in the sense that the two together are not seen though each separately is of just supraliminal strength.

In an extension of the experiments on incremental thresholds to a larger group of subjects (the original work being based on one subject only), Stiles (1953) found it necessary to postulate two additional mechanisms, called $\pi_2$ and $\pi_3$, besides the original blue, green and red mechanisms now called $\pi_1$, $\pi_4$ and $\pi_5$. The spectral sensitivities of the five postulated mechanisms are shown in Figs. 8.11 and 8.12. $\pi_3$ has nearly the same spectral sensitivity curve as $\pi_1$, but differs from it in that it determines incremental threshold only when the background is very bright. Like $\pi_1$, $\pi_4$ and $\pi_5$, it has been shown to have, at least approximately, the same spectral sensitivity curve, whether this is determined from the effects of varying the test wavelength or the background wavelength.

The $\pi_2$ mechanism determines incremental threshold only for violet flashes on very dim red, yellow, or green backgrounds. Even in this narrow range of conditions it was not usually distinguishable for one of the five subjects tested. Its spectral sensitivity to the background could be measured only at long wavelengths, and to the test field only at short, so that it is not possible to be certain that the two spectral sensitivity curves coincide.

Stiles's theory of colour discrimination has been shown to be wrong in certain details; the law of summation when a stimulus is close to threshold for two or more mechanisms differs from that assumed in the theory, and the assumption that the threshold of each mechanism is independent of the activity of the others is not exactly valid when different mechanisms are very differently active (Boynton, Das & Gardiner, 1966); but the accuracy with which this relatively simple theory explains the whole of a great body of very complex experimental data is so impressive that it is scarcely possible to doubt its general validity. It will be necessary later to discuss to what structures in the retina Stiles's 'mechanisms'

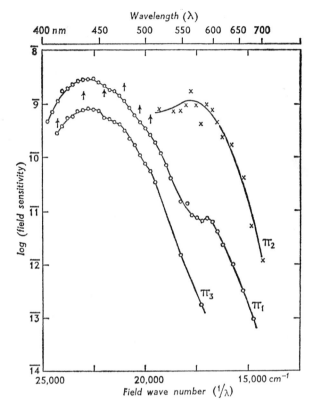

FIG. 8.11. Spectral sensitivities of the mechanisms $\pi_1$, $\pi_2$ and $\pi_3$ for changes in background wavelength. At short wavelengths, only a lower bound (shown by the arrows) can be given for the background sensitivity of $\pi_2$ (from Stiles, 1953).

correspond; but this discussion will be deferred until we have considered the evidence from spatial aspects of colour vision, from the retinal directional effect, and from some effects of adaptation to very bright lights.

## Artificial dichromacy and monochromacy produced by adapting the eye to very bright lights

Stiles's theory assumes that the raising of the threshold of a mechanism by adaptation depends only on the activity of that

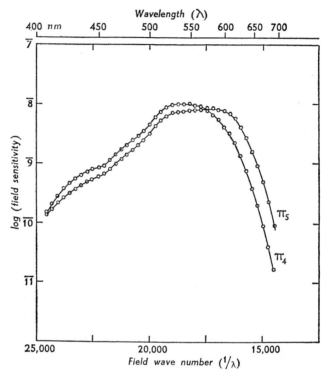

Fig. 8.12. Spectral sensitivities of the mechanisms $\pi_4$ and $\pi_5$ for changes in background wavelength (from Stiles, 1953).

mechanism itself. If this assumption is valid also for that part of adaptation which remains for some seconds or minutes after the adapting stimulus has been extinguished, it should be possible, by adaptation to very bright lights restricted to a narrow range of wavelengths, to reduce the sensitivity of one of the three mechanisms determining trichromacy so that it is entirely unaffected by stimuli that are substantially above threshold for the remaining two. In such circumstances, colour discrimination should become dichromatic; and further adaptation, depressing the sensitivity of two of the mechanisms, should abolish colour discrimination, leaving a monochromatic state.

This expectation is, to a large extent, fulfilled (Burch, 1898; Brindley, 1953); although only for one state, out of six theoretically possible on the assumption of three mechanisms, does the reduction in number of degrees of freedom of colour discrimination extend over the whole of the visible spectrum. This is the state that is found for about 40 sec after adapting to very bright violet light. During this period, all spectral lights can be matched with a mixture of red and blue primaries. The matches are satisfactorily reproducible, and the quantities of the primaries required do not vary with the degree of adaptation, provided that it is sufficient for dichromatic matches to be made at all. This state closely resembles the form of colour blindness known as tritanopia (Wright, 1952), and may for brevity be referred to as 'artificial tritanopia' (see Fig. 8.13).

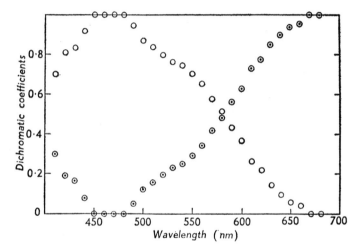

FIG. 8.13. Proportions of red (equivalent to 680 nm) and blue (480 nm) primaries required to match lights of various wavelengths in the state of artificial tritanopia. Units based on a match on 582·5 nm (from Brindley, 1953).

Artificial tritanopia can be converted by subsequent or simultaneous adaptation to very bright red or blue-green light into either of two states which will be referred to as the green and red artificial monochromacies. In these states, over a large part of the spectrum, no discrimination of colour is possible, though discrimination of brightness is but little impaired. Thus spectral sensitivity curves can be obtained by direct matching of one wavelength against another, the match, when established, being exact. Such spectral sensitivity curves are shown in Fig. 8.14. It seems almost certain that in the artificial monochromacies to which they relate, all but one of the colour-discriminating mechanisms have been rendered so insensitive by the adaptation that they contribute no information by which different wavelengths might be discriminated. The curves of Fig. 8.14 are thus presumably the spectral sensitivity curves of single channels, the red-receptive and green-receptive channels of the three-channel theory. They are compatible with estimates of these spectral sensitivities from colour matching and colour blindness (see p. 223) and with Rushton's estimates of the action spectra for bleaching erythrolabe and chlorolabe (see p. 40).

By adaptation to very bright yellow light, it is possible to produce a third state in which, over a considerable range of the

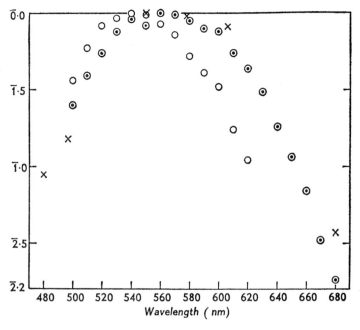

FIG. 8.14. Spectral sensitivity in energy units in the states of green artificial monochromacy (○) and red artificial monochromacy (◎). The crosses show the action spectrum for the production of a constant disturbance of colour matches, inferred from the experiment shown in Fig. 8.4 (from Brindley, 1953).

spectrum, in this case from 500 nm to the extreme violet, lights of different wavelength cannot be discriminated when their intensities are suitably adjusted; all appear of the same intensely saturated violet colour. This state, which may be called the violet artificial monochromacy, differs from the green and red artificial monochromacies in that, besides the complete absence of colour-discrimination, there is a severe defect of brightness discrimination, which makes it impossible to measure the spectral sensitivity of the eye with any useful accuracy by direct matching. Visual acuity also is very defective, the finest resolvable grating being twelve or more times larger than the finest that can be resolved with the unadapted eye.

Direct matching, as a means of measuring the spectral sensitivity curve of the eye in artificial monochromacies, has the advantage over threshold methods that the exactness of the match provides its own criterion that there is only one mechanism acting. For the violet artificial monochromacy, however, where the defect of brightness discrimination makes direct matching insufficiently accurate, the threshold method is the only one available. It has been applied by Auerbach & Wald (1954). In the range 405–492 nm the curve shown in Fig. 8.15 almost certainly corresponds to the violet artificial monochromacy.

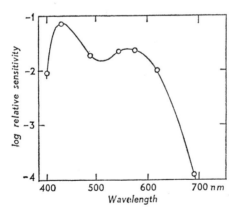

FIG. 8.15. Spectral sensitivity (in energy units) for a 1° test field centred 6° from the fixation point, measured at threshold one minute after adaptation to bright orange-red light (data of Auerbach & Wald, 1954).

### Foveal tritanopia

When very small colorimetric fields, subtending 20′ at the eye or less, are fixedly observed with the central fovea, it is found that colours can be matched with mixtures of red and blue primaries: vision is dichromatic (König, 1894; Willmer & Wright, 1945). Dichromatic coefficient curves for the spectrum for such 20′ central foveal fields are shown in Fig. 8.16. It will be seen that they

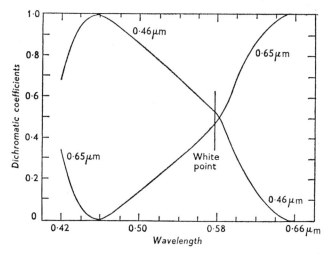

FIG. 8.16. Proportions of red (650 nm) and blue (460 nm) primaries required to match lights of various wavelengths, using a 20′ foveal field. Units based on a match on wavelength 582·5 nm (from Willmer & Wright, 1945).

are very similar to those obtained in the state of artificial tritanopia (Fig. 8.13).

This dichromacy of colour matching with very small colorimetric fields is not entirely restricted to the central fovea. Thomson & Wright (1947) showed that with fields of 15′ diameter it is still valid when these are centred 20′ or 40′ from the fixation point; but for most subjects dichromatic matches can be established more easily, and for slightly larger fields, when the fields themselves are fixated than when the fixation mark is 20′ or 40′ from them. Further evidence that the blue-blindness of the central fovea is in part a local property, and not wholly determined by the smallness of the fields and the steadiness of fixation (see also p. 244), is provided by the observation that $\pi_1$ incremental thresholds are much higher at the central fovea than half a degree or so away from it (Stiles, 1949; Brindley, 1954a; Wald, 1967).

## Spatial summation and resolution in colour-receptive pathways

It has already been noted that in the violet artificial monochromacy (a state in which the blue-sensitive channel of the three-channel theory is probably isolated) visual acuity is very defective. When a mechanism ($\pi_1$ of Stiles) which probably corresponds to the blue-sensitive channel is isolated by another method, namely by presenting a pattern in blue or violet light upon a background of longer wavelength, a similar defect of acuity is found. Figure 8.17

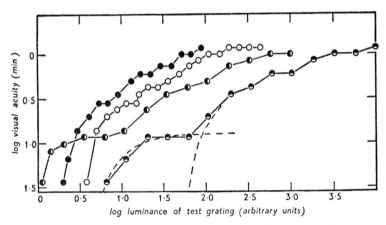

FIG. 8.17. Variation of visual acuity with luminance of test grating. ●, green grating (548 nm) on red background; ○, red (680 nm) on green; ◐, violet (454 nm) on green; ◓, violet on red. Background luminance 60 mL. The curves for violet gratings show an obvious shoulder at an acuity of about 7′ 30″, a value at which the curves for red and green gratings are descending steeply. At low luminances, violet gratings are almost certainly detected by the $\pi_1$ mechanism, and it is suggested, following Stiles's similar analysis of a comparable experiment (1949), that this mechanism cannot achieve an acuity much better than 7′ 30″ (from Brindley, 1954a).

shows the analysis of an experiment of this kind. If a grating of light and dark lines of equal width is used as test-object, the best acuity obtainable with the blue-sensitive mechanism is about 7′ 30″. With the red-sensitive and green-sensitive mechanisms it is at least eight times better.

Two hypotheses can be suggested, either of which would explain the low visual acuity in conditions which probably isolate the blue-sensitive mechanism, and would contribute towards an explanation of the tritanopia-like dichromacy found with very small foveal fields. The first is scarcity of receptors belonging to the blue-sensitive mechanism. To explain the visual acuity for a violet grating on a green or red background, the mean distance between receptors would have to correspond to about 5′. The second possible explanation of the low visual acuity associated with the blue-sensitive mechanism is that its receptors, though not scarce, are connected so that many share a common transmission line. On this hypothesis, acuity would be limited not by the receptors, but by the scarcity of their common paths. The most likely, though not the only possible, site for such a convergence is within the retina, many blue-sensitive receptors being connected to each retinal ganglion cell.

Available evidence supports the hypothesis of convergence against that of scarcity. If the blue-sensitive receptors were scarce enough to explain the low visual acuity by their scarcity alone, the incremental threshold for a very small violet spot on a green background should vary greatly from point to point on the retina, according to whether it illuminated a blue receptor or not. This is not observed (Brindley, 1954a). If, on the other hand, many blue-sensitive receptors converged on to each ganglion cell, we might expect, by analogy with the dark-adapted human peripheral retina or with the ganglion cells of the frog (see p. 86), that the product of area and intensity should be constant for threshold, up to a size of field corresponding to a substantial fraction of that area of retina to whose receptors each ganglion cell is connected. Measurements of incremental threshold as a function of size of field (Fig. 8.18) for a retinal region where sensitivity is not changing rapidly from point to point confirm this expectation. For incremental thresholds depending on the blue-sensitive mechanism, the threshold is inversely proportional to the area (Riccó's law) for fields of up to about 13′ diameter. For those depending on the red-

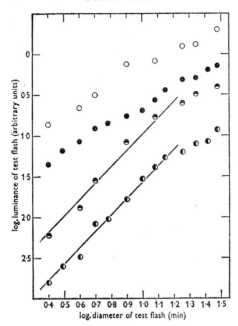

FIG. 8.18. Variation of incremental sensitivity with size of test field. Test fields centred 55' below the fixation point. ◯, red (680 nm) on green (548 nm) background; ●, green on red; ◕, violet (454 nm) on red; ◑, violet on green (from Brindley, 1954a).

sensitive and green-sensitive mechanisms, it is approximately inversely proportional to the diameter (Piper's law) over the whole of the range tested.

The above argument for convergence and against scarcity as explanation of the low visual acuity of the blue-sensitive mechanism does not oppose the view that the high $\pi_1$ threshold of the central fovea, and the constituent of foveal tritanopia that is local to the central fovea, depend on local scarcity of blue sensitive receptors there. But the general foveal mosaic of blue-sensitive receptors cannot, I think, be so coarse as, in itself, to limit acuity to $7\frac{1}{2}'$.

The fact that discriminations depending on the blue-sensitive mechanism are impaired by a decrease in size of field to a greater extent than those depending on red-sensitive and green-sensitive mechanisms provides a reason why, with very small fields, colour discrimination should approximate to a tritanopia-like dichromacy. It is, however, quantitatively quite insufficient to explain complete dichromacy with fields of 15' diameter. If the blue-sensitive mechanism were always inactive for fields of this size or less, incremental thresholds for very small violet test flashes on backgrounds of longer wavelength would depend on the red-sensitive or green-sensitive mechanisms. The very differences in spatial summation that we are considering prove that they do not. When the small fields are centrally fixated, a contributory factor in dichromacy is presumably the relatively low sensitivity of the blue-sensitive mechanism at the foveal centre (Stiles, 1949; Brindley, 1954a; Wald, 1967); but for fields separated by 20' or 40' from the foveal centre there remains a discrepancy. The observation of Willmer (1950b) that dichromatic matches cannot be made immediately on fixating a pair of very small fields, but only after a detectable delay, suggests another factor, which may perhaps be the most important. It may be that the signal sent to the brain by the blue-sensitive mechanism, though efficient at distinguishing the relative strengths of brief stimuli and at detecting sudden changes, is inefficient at distinguishing the strengths of prolonged stimuli. If so, small-field dichromacy is not only a property of the smallness and position of the fields, but also requires prolonged steady fixation. This seems to be true: McCree (1960) found that it was impossible to make dichromatic matches with small fields if they were presented to the eye for only a fraction of a second.

## Temporal resolution in colour-receptive pathways

The blue-sensitive mechanism is inferior to the red-sensitive and green-sensitive in its temporal as well as in its spatial resolving power. Brindley, du Croz and Rushton (1966) examined the flicker fusion frequency for incremental flicker of one colour on a background of a different colour. When the incremental flicker was such as would be expected to stimulate the red-sensitive or green-sensitive mechanisms (whether or not it also affected the blue-sensitive), the flicker fusion frequency rose smoothly with increasing luminance to values (for a 2° foveal field) around 55 c/sec. When it was such as would be expected to stimulate only the blue-sensitive mechanism, it rose with increasing luminance only to 18 c/sec, and at higher luminances remained constant.

There is not yet any evidence to decide whether the defective temporal resolution of the blue-sensitive mechanism is a property of receptors or of nervous pathways. Whether it is determined peripherally or centrally to the site at which electrical phosphenes are generated might be resolved by the technique of Brindley (1962a), according to the argument given on p. 174.

Part of the defective temporal resolution of the blue-sensitive mechanism, and doubtless also of its defective spatial resolution, is secondary to its poor intensity-discrimination, as Green (1969) has well illustrated by exploring the temporal resolution of the various colour mechanisms over a wide range of contrasts.

## The effects of adaptation on apparent colour

Although matches between metameric lights are stable to adaptation except at very high intensities (p. 216), much smaller degrees of adaptation suffice to cause conspicuous changes in the appearance of colours. These successive contrast phenomena, or 'after-images' in the wide sense of that term, have been investigated quantitatively by several writers of the late nineteenth century (see p. 151), and more recently by Wright (1934), Walters (1942) and MacAdam (1949a, 1956), by adjusting stimuli until the same quality of sensation was obtained from two areas of the retina which were in different adaptational states. The early writers and MacAdam placed the two stimuli on different parts of the same retina; Wright and Walters placed them on non-corresponding parts of the left and right retinas. Wright and Walters attempted to use the results that they obtained to estimate the position in the colour triangle of the hypothetical 'fundamental stimuli', each of which stimulates one only of the three channels of the three-channel hypothesis. The assumptions required in order to do this were not explicitly stated by Wright or by Walters, but it is clear that they amount to the following:

1. The effect of a change of adaptational state on the sensitivity of any one of the three channels which determine trichromacy is to alter it by a factor that is independent of wavelength.
2. The sensation produced by any stimulus is absolutely determined by the extents to which it stimulates the three channels; if in an adaptational state $A$ the three channels are

I

respectively $x$, $y$ and $z$ times more sensitive than in an adaptational state $B$, then a stimulus made up of amounts $r$, $g$ and $b$ of the three 'fundamental stimuli' applied to an area of retina in state $A$ will produce the same sensation as a stimulus made up of amounts $xr$, $yg$ and $zb$ applied to an area of retina in state $B$.

Provided that very bright adapting lights are excluded, the first of these assumptions is strongly supported by the stability of colour matches to adaptation. The second assumption is much more uncertain. There is no theoretical reason why it should be true, and the existence of after-images visible against a background of darkness is direct proof that it cannot be exactly so. Broadly, there are two factors that would tend to make it false: the receptors may discharge signals in darkness, as well as when stimulated by light, the nature of these signals being affected by adaptation; and some component of adaptation may (as for scotopic vision it unquestionably does) occur centrally to the receptors, in the neural layers of the retina or the brain. Here the number of channels carrying information about colour may be more than three, and there is no reason why their spectral sensitivities should be stable to adaptation.

The second of these factors affects all published experiments, and, it seems, all possible ones. The effect of adaptation on activity of the receptors in darkness is relevant to the nineteenth-century experiments and those of MacAdam, but is eliminated in Wright's work, because within a region of the visual field including that in which both test stimuli fell, one retina was uniformly affected and the other uniformly unaffected by the pre-adaptation.

From the two assumptions stated above as necessary for deriving the fundamental stimuli, we can make a deduction capable of experimental test. If a light $A$ falling on one area of the retina produces the same quality of sensation as a light $B$ falling on another area, and a light $C$ on the first the same quality of sensation as a light $D$ on the second, then $(A+C)$ falling on the first must produce the same quality of sensation as $(B+D)$ falling on the second. This is the 'law of coefficients', first expressed in this clear explicit form by von Kries (1905), though several earlier writers (to whom von Kries refers) had made obscure statements which may have been intended to convey the same meaning.

If the 'law of coefficients' were exactly true, it would not prove the validity of the two assumptions required in the derivation of the fundamental stimuli, but would furnish good presumptive evidence in their favour. In fact, as the experiments of Walters clearly show, it is false, though under a limited range of conditions it can be a fair approximation to the truth. Its falsity proves that the assumptions required in the derivation of the fundamental stimuli cannot be exactly true under the conditions for which Wright and Walters used them. It is difficult to assess to what extent the estimates of the fundamental stimuli derived from experiments on adaptation should nevertheless be approximately valid. My opinion is that they are probably fairly good approximations; but this opinion is based more upon their similarity to estimates from other assumptions and experiments than upon the compelling power of the arguments from adaptation.

MacAdam (1949a, 1956) argued from the results which he obtained in experiments essentially similar to those of Wright in favour of a six-receptor hypothesis. Like Walters, MacAdam found deviations from the 'law of coefficients'. However, there is no difficulty in explaining these deviations within a three-receptor or three-pigment hypothesis, if we suppose that some component of adaptation occurs centrally to the receptors; hence there is no justification for preferring a complex six-receptor hypothesis, which predicts nothing whilst it remains general, but can be specialized to fit almost any experimental result, to a relatively simple three-receptor or three-pigment hypothesis of wide predictive power.

None of the published experiments on the effects of adaptation on apparent colour belongs properly to Class A, because in all of them the subject can know from the position of each sensation in the visual field what was the adaptational state of the region of retina upon whose stimulation it depended: there is no true indistinguishability. Probably a Class A experiment could be done if the two eyes were put into different adaptational states, and stimuli presented briefly to one or other eye without giving the subject any external clue informing him which eye had been stimulated. The practicability of such an experiment depends on discovering conditions where a subject cannot distinguish the stimulated eye by an internal clue. Whether such conditions exist is controversial, but my own unpublished experiments and the work of Templeton & Green (1968) and Enoch, Goldmann & Sunga (1969) convince me that they do, at least for some subjects.

The true Class A condition shares with that of Wright's experiments the advantage that any effects of adaptation on the resting discharge of colour-receptive channels that may occur in addition to effects on sensitivity are equal for the two stimuli to be compared. Wright's experiments were interpreted above by the third of the methods indicated in Chapter 5 for dealing with Class B observations, namely by introducing two special assumptions, one of which is a psycho-physical linking hypothesis and unfortunately turns out to be false. It may be more useful to interpret the experiments according to the second of the methods of Chapter 5, namely as a means of extrapolating to closely related but technically more difficult Class A experiments which can in principle be used directly to test physiological hypotheses, though the appropriate hypotheses have not yet been developed.

## The directional selectivity of the retina (Stiles–Crawford effect)

### The fovea

Stiles & Crawford (1933) showed that the sensitivity of the eye varies with the direction of incidence of light upon the retina, i.e. with whether it has come through the centre or the periphery of the pupil. Only a small fraction of this directional selectivity can be accounted for by differences in losses of light within the eye; much the greater part of it must be a property of the retinal receptors.

The directional effect can be shown at foveal threshold or by matching, and for all the kinds of foveal incremental thresholds, and it is clear that if there are three kinds of foveal cones, all of them must be directionally selective at all wavelengths. However, the amount of directional selectivity at a given wavelength differs for different kinds of cones, as Stiles (1939) showed for the green-sensitive and blue-sensitive incremental threshold mechanisms ($\pi_4$ and $\pi_1$); and the directional selectivity of at least one of the kinds of foveal cone must vary with wavelength, since colour matches valid for one angle of incidence are not valid for another (Brindley, 1953; see p. 249). Enoch & Stiles (1961) examined in detail the effect of angle of incidence on the chromaticity of monochromatic lights. By making plausible assumptions about the positions of the 'fundamental stimuli' in the colour triangle (see p. 223) they estimated the variation of directional selectivity with wavelength for each of the three presumed kinds of foveal cone. For the red- and green-sensitive cones, directional selectivity is low at wavelengths where sensitivity is high, and high where sensitivity is low. For the blue-sensitive cones, there is no such relation.

*The effects of direction of incidence on colour matches.* Figure 8.19 shows the way in which the amounts of red and green primaries required to match monochromatic lights of wavelengths from 560 to 780 nm alter when all the light is sent into the eye through the periphery of the pupil instead of its centre. The disturbance of colour matches strikingly resembles that produced by adaptation to very bright lights (Fig. 8.3). The self-screening hypothesis discussed on pp. 213 and 221 immediately explains this resemblance. If the self-screening hypothesis is false, the resemblance must be a mere coincidence. The view that it is not mere coincidence is strengthened by the fact that adaptation to very bright light abolishes the effect of obliquity on the hue of spectral yellows (Brindley, 1953), and also abolishes the minimum in the Stiles–Crawford effect at about 580 nm (Walraven, 1966).

### Extrafoveal retina

If the sensitivity of extrafoveal retina is tested at absolute threshold with light of wavelength less than 600 nm, i.e. under conditions where it is almost certainly determined solely by rods, it is found to be nearly independent of the direction of incidence of the light (Stiles, 1939). At longer wavelengths, the extrafoveal retina shows substantial directional selectivity even at absolute threshold. But this directional selectivity is due to cones, not rods, as Flamant & Stiles (1948) showed by measuring the spectral sensitivities of extrafoveal retina for light entering through the centre and the periphery of the pupil, and comparing them with the energies of adapting fields of different wavelengths required to raise the threshold for blue-green light by a constant factor. The rods, it seems, retain their lack of directional selectivity at all wavelengths.

Extrafoveal cones are more directionally selective than central foveal cones (Westheimer, 1967b). Central foveal cones are very different in shape from those of the rest of the retina and their outer segments resemble those of rods. It is thus not surprising that in directional selectivity they are intermediate between rods and extrafoveal cones; one might rather be surprised at the fact that they are, in this respect, a good deal closer to extrafoveal cones than they are to rods.

### Light entering the retina from behind or from the side

The most extreme way of varying the direction of incidence of light on the receptors is to pass it through the sclera, so that it enters the rods and cones through their outer ends. Brindley & Rushton (1959) did this, as a test of the possibility that the differences between the spectral sensitivities of different receptors that are required to account for colour vision might be due not (or not only) to differences in the absorption spectra of the pigments that they contain, but to some mechanism that selects the light admitted to the place where the pigment is situated. This mechanism might be a light-absorbing filter, such as birds are known to possess, or something analogous to a waveguide. Any of these devices would be expected to be ineffective, or at least differently effective, for light entering the receptors from their outer ends. Thus monochromatic light sent into the retina trans-sclerally should stimulate the receptors in different proportions from light of the same wavelength sent in through the pupil, and so should look different in colour; on the extreme hypothesis of a single receptive pigment with selectors in front of it, the apparent colour of a light sent in through the sclera should be independent of its wavelength. This is far from the truth. In fact, green light sent through the sclera looks green, yellow looks yellow, and red looks red, each differing only very slightly in appearance from the same light sent in through the pupil.

A waveguide action of cones might perhaps fail to discriminate between light coming along the axis from in front or behind; but it would necessarily discriminate between axial light and light coming from the side. One might hope to examine the sensitivity of foveal cones to laterally incident light by strongly illuminating the optic disc. It is easy to verify that this causes a sensation of light over the whole of the visual field. I have done experiments in which masking light was put continuously in the whole extramacular field, and strong monochromatic illumination of the disc was alternated with a weak controllable comparison light seen directly by the macular region. The experiments show that the sensation in the central part of the visual field when the disc is strongly illuminated is indeed due to stimulation of the cones of the macular region by light reflected from the disc. The sensation matches that from monochromatic light of a wavelength only slightly different (up to 10 nm shorter in the green, up to 10 nm longer in the orange and near red) put through the pupil. But the interpretation is less satisfactory than one might hope, because the path of the light from disc to macula is not mainly direct, as is proved by the fact that a given luminous flux (in photopic photometric units) falling on the disc produces a much stronger sensation in the centre of the visual field if it is blue or violet than if it is green or yellow. The reflectance of the disc is not higher but lower in the blue and violet; thus the only plausible explanation is that the light mainly follows paths that involve multiple reflexion or scattering. If so, the direction of incidence of the scattered light on the cones of the macular region is quite uncertain.

A good experiment on the sensitivity of rods and cones to very oblique light could probably be done on a subject with an old patch of choroiditis or a congenital coloboma of the choroid. Such a lesion would probably be larger and have higher reflectance than the optic disc, and part at least of it would be less extremely oblique in relation to the macular retina.

## The usefulness of directional selectivity

The optical mechanism of directional selectivity is, as we shall see (p. 252), rather obscure, but many of the rough models that have been proposed for it imply that sensitivity to obliquely incident light is not merely thrown away, but traded for higher sensitivity to directly incident light. The photosensitivity of cone pigments *in situ* (Brindley, 1955*b*, Rushton, 1963*c*, 1965*b*) is substantially greater than that of rhodopsin *in situ*, and this in turn is slightly greater than that of either rhodopsin or iodopsin in solution, when allowance has been made for transmission losses and orientation of the chromophores. There is thus little doubt that funnelling occurs: the illumination of the rhodopsin in the rods is at least a little greater, and that of the visual pigments in the cones substantially greater, than the illumination of the retina as a whole.

It is hard to see that the relative insensitivity of cones to light that comes from the outer part of the pupil can be to more than a very small extent either useful or harmful, since in ordinary life they rarely receive such light. It is natural to suggest that the important achievement of evolution is enhanced sensitivity to centrally incident light, and that the depressed sensitivity to obliquely incident light is a functionally almost neutral by-product. Rods, unlike cones, are used chiefly when the pupil is dilated, and thus need to be sensitive to light from its margins as well as its centre. Presumably it is the freedom that the cones have to sacrifice sensitivity to oblique light that has allowed them to achieve better funnelling of direct light than rods.

Properties of the dark deformation phosphene (see p. 159) provide an illustration of how sensitive the funnelling mechanism of foveal cones is to mechanical distortion of the retina. The disturbance of funnelling by such distortion does not seem to represent a mere tilting of the receptors so that they point in another direction but preserve their selectivity, since the decreased sensitivity to directly incident light is not accompanied by increased sensitivity to light from some peripheral part of the pupil.

## The mechanism of directional selectivity

Early attempts to explain the directional selectivity of cones used optical principles that were properly applicable only to systems large compared with the wavelength of light. The authors doubtless knew (and in some cases said) that their hypotheses could for structures as small as cones only be rough approximations, but at that time no better methods were known. The best of the early ideas was that of Wright & Nelson (1936), who suggested that the outer segment of the cone had a higher refractive index than its environment, so that light that entered it along the axis was trapped inside by repeated total internal reflection, but oblique light escaped. The high refractive index of outer segments was later shown experimentally by Sidman (1957; see p. 25).

Toraldo di Francia (1948) pointed out that the proper kind of theory for analysing the behaviour of cones was that of waveguides. Unfortunately this theory is not simple even when the shape of the waveguide and the refractive indices of its parts are known exactly, and when the shape is simple. If the shape is complex and inexactly known, and the refractive indices still more uncertain, it becomes difficult to use waveguide theory helpfully. A potentially useful device (Jean & O'Brien, 1949) is to make a large-scale model of the structure and test it with radio waves. But a proper model of this kind ought to contain a radio analogue of the visual pigment. Such an analogue is practicable, but I have read of no experiment on a model containing one.

The directional selectivity of a waveguide always depends on the wavelength of the radiation, and sometimes a small change in the wavelength may change the directional selectivity greatly. This property is important for visual physiology in two ways. First, it must contribute to explaining the effects of wavelength on the directional selectivity of single classes of cones; secondly, it implies that the spectral sensitivity of a cone need not agree with the photo-sensitivity spectrum that its receptive pigment would have if isolated in solution. The experiment of Brindley & Rushton (1959) excludes extreme hypotheses, such as that of a single pigment contained in three different kinds of waveguide, but is compatible with small wavelength-dependences in the waveguide action of cones. Good evidence is scarce, but what there is suggests that the wavelength-dependences are too small to introduce substantial discrepancies between the photosensitivity spectra of cone pigments in and outside cones. The experiment of Brindley and Rushton and its extension (pp. 250–251) to light scattered from the optic disc to the fovea point in this direction. The close correlation of the scotopic spectral sensitivity of the eye with the photosensitivity spectrum of rhodopsin is weak evidence or none, since it could be argued that waveguide action is slight in rods. But the effects on matches between near red and far red lights of putting the light through the periphery of the pupil (Fig. 8.19) are very pertinent. The directional selectivity of the red-sensitive cones in the red part of the spectrum is very high. Yet Figure 8.19 shows it to be practically independent of wavelength between 660 and 780 nm.

Whatever the wavelength-dependence of the waveguide action itself, any funnelling must, through change in self-screening, introduce an effect of direction of incidence on spectral sensitivity if the optical density of pigment through which the light passes is at some wavelengths high for the optimal direction of incidence; for the density will in general be lower for non-optimal directions. Pigment densities in the rods of many species are known to be high enough for such self-screening effects to be significant (see p. 27) though the effects have not been detected in man. For cones, the

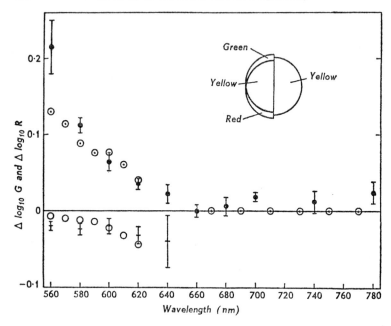

Fig. 8.19. Changes produced by a 3 mm decentring of the eye on the artificial pupil in the amounts of red (680 nm; ↨) and green (550 nm; ↧) primaries required to match test lights of various wavelengths. ⊙ and ○ show the values calculated on the hypothesis that the red receptors contain a high density of pigment, and that light coming from the margin of the pupil passes through only a part of this (from Brindley, 1953).

densities directly observed by spectrophotometry in the isolated retina (Brindley, 1963a; Marks, Dobelle & MacNichol, 1964; Brown & Wald, 1964) are much too low, and those inferred from ophthalmoscopic densitometry (e.g. Rushton, 1963b) slightly too low, to permit an explanation of the results of Figures 8.3–8.6 and 8.19 in terms of self-screening. But all the objective estimates are extremely fallible (see p. 44); for example, in the spectrophotometry of single cones the postmortem distortion of the retina must almost certainly abolish funnelling, and measurements on deformation phosphenes (see p. 160) prove that this reduces the sensitivity of cones by a factor of about 6 and suggest that it may reduce the mean density of cone pigment through which the rays pass by almost as large a factor. Stray light, unless corrected for (as

Rushton has attempted with some success) may introduce even larger errors in the objective estimates.

## Relation of the mechanisms determining incremental thresholds to the three channels responsible for trichromacy

The simplest hypothesis that we can make about the mechanisms which determine incremental thresholds is that they correspond directly to the channels responsible for trichromacy. In accepting this hypothesis there is the immediate difficulty that five mechanisms are required to account for Stiles's data on incremental thresholds. Of these five, $\pi_3$, the blue mechanism which comes into action at high intensity, would not provide information independent of that given by its low-intensity counterpart $\pi_1$, since it has almost the same spectral sensitivity curve; but $\pi_1$, $\pi_2$, $\pi_4$, and $\pi_5$ appear from their spectral sensitivity curves to provide mutually independent information, so that they should determine a tetrachromacy and not a trichromacy. There are two ways in which the facts and theory might be reconciled without entirely rejecting the simple hypothesis of a correspondence between the mechanisms determining incremental thresholds and those determining trichromacy. The spectral sensitivities of the four incremental threshold mechanisms might be related, to a close enough approximation for departures never to be supraliminal, by a linear equation with coefficients independent of wavelength; or one of the four might not be a channel in the same anatomical sense as the other three, but represent some kind of interaction between them which allows certain test flashes to be seen which without the interaction would have been subliminal. On the former alternative the colour-matching properties can be deduced from the spectral sensitivities of any three of the mechanisms, unless the one omitted does not appear in the linear equation; we should naturally choose the three whose spectral sensitivities have been most accurately determined, namely $\pi_1$, $\pi_4$ and $\pi_5$. On the latter alternative, the implications for colour matching depend on which mechanism has the subordinate status of arising merely from an interaction between the other three. The properties to be expected of a mechanism of this kind are analogous to those of a channel in the third form of the three-channel hypothesis: there is no necessary reason why its spectral sensitivity curve should have the same shape at different intensities, or be the same when measured by varying the test wavelength as by varying the background wavelength, and some ground for expecting that it should not. For a mechanism that does not represent an interaction, on the other hand, the spectral sensitivity curve must, if there is only one photosensitive pigment concerned, be an invariant property of that pigment. These considerations provide a reason for suggesting that if one of the 'mechanisms' determining incremental thresholds represents merely an interaction between the others, it is likely to be $\pi_2$, whose spectral sensitivities to test lights and to adapting fields have not been measured in corresponding parts of the spectrum and may differ, rather than $\pi_1$, $\pi_4$ or $\pi_5$, for which the two sensitivities are known to be in fair agreement in the spectral regions where both can be measured.

Stiles (1953) has examined the colour-matching properties of a hypothetical eye which gains all its information from the mechanisms $\pi_1$, $\pi_4$ and $\pi_5$. He first compares the spectral sensitivity curves of these mechanisms with the best-fitting linear combinations of the C.I.E. colour mixture curves and of Judd's revision of these curves. The agreement, especially when Judd's revised curves are used, is close, but there remains a small but significant discrepancy which suffices to exclude the hypothesis that the colour-matching properties of Stiles's five subjects are determined by three mechanisms with the spectral sensitivities of $\pi_1$, $\pi_4$ and $\pi_5$, unless Stiles's subjects are substantially different in their colour-matching properties from Wright's, or unless there is considerable inaccuracy in inferring the spectral sensitivities of the incremental threshold mechanisms. A substantial difference in the colour-matching properties of the two groups of subjects is unlikely, in view of the close agreement in this respect between Wright's subjects and those studied by other investigators (see p. 203). The possible error in inferring the spectral sensitivities of $\pi_1$, $\pi_4$ and $\pi_5$ from the measurements on incremental thresholds is difficult to assess from Stiles's published curves, but it seems that it could perhaps be large enough to account for the discrepancy.

It is evident that the simple hypothesis that the mechanisms determining incremental thresholds correspond directly to the channels which determine trichromacy has to be a good deal strained to fit all the facts, and probably requires some modification. There is insufficient evidence to make it profitable to discuss the possible modifications in detail, but one suggestion may be worth making, because it can be tested by experiment. The shoulder on the spectral sensitivity curve of $\pi_1$ around wavelength 600 nm is a somewhat unlikely feature of an absorption spectrum of a single pigment, and it is absent in the spectral sensitivity curve of the $\pi_3$ mechanism, which in other respects is so similar that it might be supposed to depend on the same receptive pigment. The suggestion is that the shoulder on the curve as measured by the incremental threshold technique is due to suppression or masking of the signal transmitted from the blue-sensitive receptors by the much greater activity, under the conditions in which the field sensitivity of the $\pi_1$ mechanism is measured at long wavelengths, of the red-sensitive and green-sensitive ones. Such masking might occur in any part of the visual pathway, from the receptors to the cortex. An experimental test for it is to add, in an experiment where the incremental threshold for blue light on a red or orange background is being measured, a little blue light to the background. On the original hypothesis of independent mechanisms, this must raise the threshold by exactly as much as does the addition of an amount of red or orange light equivalent to it on the spectral sensitivity curve of $\pi_1$. On the new hypothesis there is no reason why the blue supplement to the background should have the same effect as the 'equivalent' red or orange supplement; it is likely to be generally much less effective.

This interaction suggested to explain the shoulder on the $\pi_1$ curve differs from the one proposed on p. 255 to reconcile the trichromacy of colour matching with the presence of more than three increment threshold mechanisms. The latter kind of interaction makes visible certain test flashes which without it would be subliminal; the former makes invisible certain flashes which, if the mechanisms were completely independent, could be seen.

## Can the blue receptors be rods?

It has been suggested by several writers from König (1894) to Willmer (1946) that trichromatic colour discrimination depends on only two kinds of cones, the third degree of freedom being provided by rods, which, still using rhodopsin as their receptive pigment, act as the blue-sensitive channel of the three-channel hypothesis. The principal evidence in favour of this suggestion is the dichromacy of colour discrimination with small foveal fields, which illuminate no rods (see p. 240). The quantitative agreement between the size of field required to produce dichromacy and the size of the foveal region which histologically contains no rods is neither good enough to support the hypothesis strongly, nor bad enough to disprove it. The diameter of the foveal region entirely free from rods is said by Polyak (1941) to be 500 to 600 μm (1° 42' to 2° 3'), and by Østerberg (1935) to be 260 μm (53'). Rochon-Duvigneaud (1943) gives the very low value of 150 to 200 μm (31' to 41'), but this is uncorrected for shrinkage. If Østerberg's correction for shrinkage is applied to it, it comes into approximate agreement with his figure. Dichromatic colour matches can be made by most subjects on a bipartite circular foveal field of 20' diameter, but not on one of 30' diameter. The failure to make dichromatic matches on a 30' field is difficult to reconcile with the histological data on the hypothesis of two kinds of cones and one of rods, but is not altogether inconsistent with them if the lowest histological figures are taken and if it is assumed that subjects do not fixate the test fields very accurately.

The idea of rods as the blue-sensitive receptors might be thought to gain in attractiveness from the demonstration (see pp. 241–245) that the blue-sensitive mechanism differs from the red- and green-sensitive mechanisms and resembles the rod system in spatial and temporal resolving power and in spatial summation. Nevertheless, it is certainly false, as the following argument proves. If receptors are of only three kinds, it is obvious that any two lights which affect all three kinds alike must match; and the converse of this, namely that any two lights which match must affect all three kinds of receptors alike, can be proved from generally acceptable assumptions (Brindley, 1957c). It is well established (König, 1896; von Kries, 1896; Tschermak, 1898; Stiles, 1955) that if, in a trichromatic colour match established for a field large enough or far enough from the fixation point to illuminate a substantial number of rods, all constituents are attenuated in the same proportion nearly or quite to the brightness-level at which colour discrimination is lost, the match may cease to hold. This breakdown cannot be attributed to a change in the spectral sensitivities of receptors arising from photolysis of a screening pigment, for it occurs at light intensities far below those at which substantial fractions of any retinal pigments are bleached (Brindley, 1955b; Rushton, 1956a, 1963b). If there are only three kinds of receptors illuminated, including the rods, then the two lights which match at the relatively high intensity must affect all receptors, including the rods, alike, and hence must still match when all constituents are attenuated in the same proportion. Since they do not, the hypothesis is contradicted.

The above argument establishes that the blue-sensitive receptors are not the ordinary rods that contain rhodopsin, and proves that the retina contains at least four visual pigments including rhodopsin. It does not exclude the notion that the blue-sensitive receptors may be rods by anatomical criteria, but contain cyanolabe instead of rhodopsin. Anatomical evidence and the spectrophotometric observations on single receptors described on pp. 44–46 oppose this notion, but perhaps do not quite disprove it.

# BIBLIOGRAPHY AND INDEX OF
# AUTHORS

*Alphabetical order is determined by the Anglo-American Cataloguing Rules (1967). Thus, for example E. De Renzi and R. De Valois are indexed under D because they are Italian and American respectively, and E. De Robertis because, although he is Argentinian, his name is not of Spanish origin and he himself indexes it under D. On the other hand, H. de Lange is indexed under L because he was of Dutch nationality and language, and J. da Fonseca under F because he is of Portuguese nationality and language. S. Ramón y Cajal appears under C (with a cross-reference from R) because this was his own practice, despite the contrary Spanish custom.*

*Numbers in square brackets at the end of an entry indicate the pages of text on which reference is made to it.*

Abelsdorff, G. (1895). Über die Erkennbarkeit des Sehpurpurs von Abramis brama mit Hülfe des Augenspiegels. *S. B. Akad. Wiss. Berlin*, **1895**, 325–329. [28]
— (1898). Physiologische Beobachtungen am Auge der Krokodile. *Arch. Anat. Physiol. Lpz.*, **1898**, 156–167. [28]
Abelsdorff, G. & Nagel, W. (1904). Über die Wahrnehmung der Blutbewegung in den Netzhautkapillaren. *Z. Psychol. Physiol. Sinnesorg*, **34**, 291–299. [142]
Abelsdorff, G. see also Köttgen, E.
Abney, Sir William de W. (1913). *Researches in colour vision and the trichromatic theory*. Longmans, London. [145, 154, 203]
Abney, W. de W. & Festing, E. R. (1886). Colour photometry. *Phil. Trans.*, **177**, 423–456. [144]
Abrahamson, E. W. see Wulff, V. J., Ostroy, S. E., Poincelot, R. P.
Adams, R. G. see Wulff, V. J.
Adamük, E. (1870). Über die Innervation der Augenbewegungen. *Zbl. med. Wiss.*, **8**, 65–67. [93]
Ades, H. W. & Raab, D. H. (1949). Effects of preoccipital and temporal decortication on learned visual discrimination in monkeys. *J. Neurophysiol.*, **2**, 101–108. [130]
Ades, H. W. see also Marshall, W. H.
Adler, H. E. see Brown, J. L.

Adorjani, C. see Denney, D.

Adrian, E. D. (1945). The electric response of the human eye. *J. Physiol.*, **104**, 84–104.                                                                      [73, 74]

Adrian, E. D. & Matthews, B. H. C. (1934). The Berger rhythm: potential changes from the occipital lobes in man. *Brain*, **57**, 355–385.          [121]

Adrian, E. D. & Matthews, R. (1927). The discharge of impulses in the optic nerve and its relation to the electric changes in the retina. *J. Physiol.*, **63**, 378–414.                                                                     [63, 81]

— (1928*a*). The processes involved in retinal excitation. *J. Physiol.*, **64**, 279–301.                                                                                 [81]

— (1928*b*). The interaction of retinal neurones. *J. Physiol.*, **65**, 273–298.
                                                                                          [81]

Aguilar, M. & Stiles, W. S. (1954). Saturation of the rod mechanism of the retina at high levels of stimulation. *Optica Acta*, **1**, 59–65.     [175]

Ajmone Marsan, C. see Widén, L.

Alpern, M. & Campbell, F. W. (1963). The behaviour of the pupil during dark-adaptation. *J. Physiol.*, **165**, 5P–7P.                        [182]

Alpern, M. & David, H. (1959). The additivity of contrast in the human eye. *J. gen. Physiol.*, **43**, 109–126.                                       [152]

Alpern, M., Lee, G. B. & Spivey, B. E. (1965). $\pi_1$ cone monochromatism. *Arch. Ophthal.*, **74**, 334–337.                                              [227]

Alpern, M., Rushton, W. A. H. & Torii, S. (1970). The size of rod signals. *J. Physiol.*, **206**, 193–208.                                                 [56]

Alpern, M. see also Falls, H. F.

Anderson, K. V. & Symmes, D. (1969). The superior colliculus and higher visual functions in the monkey. *Brain Res.*, **13**, 37–52.             [93]

Andrews, D. P. see Teller, D. Y.

Angel, A., Magni, F. & Strata, P. (1965). Evidence for presynaptic inhibition in the lateral geniculate body. *Nature*, **208**, 495–496. [102]

Anstis, S. M. (1967). Visual adaptation to gradual change of intensity. *Science*, **155**, 710–712.                                                          [148]

Apter, J. T. (1946). Eye movements following strychninization of the superior colliculus of cats. *J. Neurophysiol.*, **9**, 73–86.           [93]

Appelmans, M. see Missotten, L.

Arden, G. B. (1954). The dark reactions in visual cell suspensions. *J. Physiol.*, **123**, 386–395.                                                       [27]

Arden, G. B., Bridges, C. D. B., Ikeda, H. & Siegel, I. M. (1966). Isolation of a new fast component of the early receptor potential. *J. Physiol.*, **186**, 123P–124P.                                                            [31]

Arden, G. B. & Ernst, W. (1969). Mechanisms of current production found in pigeon cones but not in pigeon or rat rods. *Nature*, **223**, 528–531.
                                                                                          [49]

Arden, G. B. & Miller, G. L. (1968). Generation of the vertebrate early receptor potential and its relation to rhodopsin chemistry. *Nature*, **218**, 646–649.                                                                      [32]

Arden, G. B. & Weale, R. A. (1954). Nervous mechanisms and dark-adaptation. *J. Physiol.*, **125**, 417–426.                                  [178, 179]

Arey, L. B. (1916). The function of the efferent fibers of the optic nerve of fishes. *J. comp. Neurol.*, **26**, 213–245. [107]

Aristotle. *Parva Naturalia*, 459b. [145]

Armington, J. C., Johnson, E. P. & Riggs, L. A. (1952). The scotopic A-wave in the electrical response of the human retina. *J. Physiol.*, **118**, 289–298. [74]

Arnott, G. P. see Davis, R. J.

Asher, H. (1951). The electroretinogram of the blind spot. *J. Physiol.*, **112**, 40P. [64]

Aubert, H. (1865). *Physiologie der Netzhaut.* Morgenstern, Breslau. [166, 203]

Auerbach, E. & Wald, G. (1954). Identification of a violet receptor in human color vision. *Science*, **120**, 401–405. [239]

Ayres, W. C. (1882). Zum chemischen Verhalten des Sehpurpurs. *Untersuchungen physiol. Inst. Univ. Heidelberg*, **2**, 444–447. [4]

Ayres, W. C. & Kühne, W. (1882). Über Regeneration des Sehpurpurs beim Säugethiere. *Untersuchungen physiol. Inst. Univ. Heidelberg*, **2**, 215–240. [1, 180]

Baker, H. D. (1955). Some direct comparisons between light and dark adaptation. *J. opt. Soc. Amer.*, **45**, 839–844. [185]

Baker, H. D., Fulton, A. B. & Rushton, W. A. H. (1969). Dark adaptation: free opsin or regeneration rate? *J. Physiol.*, **204**, 124P–125P. [185]

Baker, H. D. see also Rushton, W. A. H.

Ball, S., Collins, F. D., Dalvi, P. D. & Morton, R. A. (1949). Reactions of retinene$_1$ with amino compounds. *Biochem. J.*, **45**, 304–307. [19]

Bangham, A. D. see Bonting, S. L.

Barlow, H. B. (1953*a*). Action potentials from the frog's retina. *J. Physiol.*, **119**, 56–68. [82]

— (1953*b*). Summation and inhibition in the frog's retina. *J. Physiol.*, **119**, 69–88. [86, 88, 177]

— (1956). Retinal noise and absolute threshold. *J. opt. Soc. Amer.*, **46**, 634–639. [190]

— (1957). Increment thresholds at low intensities considered as signal-noise discriminations. *J. Physiol.*, **136**, 469–488. [175, 197]

— (1958). Temporal and spatial summation in human vision at different background intensities. *J. Physiol.* **141**, 337–350. [56, 164, 169, 171, 172, 176, 196]

— (1962). Measurements of the quantum efficiency of discrimination in human scotopic vision. *J. Physiol.*, **160**, 169–188. [197]

— (1963). Slippage of contact lenses and other artefacts in relation to fading and regeneration of supposedly stable retinal images. *Q. J. exp. Psychol.* **15**, 36–51. [148]

— (1964). Dark-adaptation: a new hypothesis. *Vision Res.*, **4**, 47–57. [181]

Barlow, H. B., Blakemore, C. & Pettigrew, J. D. (1967). The neural mechanism of binocular depth discrimination. *J. Physiol.*, **193**, 327–342. [119]

Barlow, H. B. & Brindley, G. S. (1963). Inter-ocular transfer of movement after-effects during pressure blinding of the stimulated eye. *Nature*, **200**, 1346–1347. [145]

Barlow, H. B., FitzHugh, R. & Kuffler, S. W. (1957). Change of organization in the receptive fields of the cat's retina during dark adaptation. *J. Physiol.*, **137**, 338–354. [87]

Barlow, H. B. & Hill, R. M. (1963). Selective sensitivity to direction of movement in ganglion cells of the rabbit retina. *Science*, **139**, 412–414. [88]

Barlow, H. B. & Levick, W. R. (1965). The mechanism of directionally selective units in rabbit's retina. *J. Physiol.*, **178**, 477–504. [88]

— (1969a). Three factors limiting the reliable detection of light by retinal ganglion cells of the cat. *J. Physiol.*, **200**, 1–24. [81]

— (1969b). Changes in the maintained discharge with adaptation level in the cat retina. *J. Physiol.*, **202**, 699–718. [82]

Barlow, H. B. & Sparrock, J. M. B. (1964). The role of after-images in dark adaptation. *Science*, **144**, 1309–1314. [151, 182]

Barlow, H. B. see also Kuffler, S. W., Teller, D. Y.

Barry, J. see Pribram, H. B.

Bartlett, N. R. see Johnson, E. P.

Bates, J. A. V. & Ettlinger, G. (1960). Posterior biparietal ablations in the monkey. *Arch. Neurol.*, **3**, 177–192. [131]

Battersby, W. S. see Teuber, H. -L.

Bauereisen, E. see Küchler, G., Sickel, W.

Baumgardt, E. (1953). Seuils visuels et quanta de lumière. Précisions. *Année psychol.*, **53**, 431–441. [195]

— (1955). Un cas d'achromatopsie atypique. *J. Physiol. (Paris)*, **47**, 83–87. [227]

von Baumgarten, R. see Jung, R.

Baumgartner, G. see Denney, D., Jung, R.

Beams, J. W. (1934). The eye as an integrator of short light flashes. *J. opt. Soc. Amer.*, **25**, 48. [171]

Bechterew, W. (1883). Über den Verlauf der die Pupille verengenden Nervenfasern im Gehirn und über die Localisation eines Centrums für die Iris und Contraction der Augenmuskeln. *Pflügers Arch. ges. Physiol.*, **31**, 60–87. [94]

Beeler, G. B. (1967). Visual threshold changes resulting from spontaneous saccadic eye movements. *Vision Res.*, **7**, 769–775. [186]

Bender, M. B. see Pasik, T., Teuber, H. -L.

Benham, C. E. (1894). The earliest description of his top appears in an anonymous note in *Nature*, **51**, 113. [154]

Berger, H. (1929). Über das Elektrenkephalogramm des Menschen. *Arch. Psychiat. Nervenkr.*, **87**, 527–570. [121]

Berkley, M., Wolf, E. & Glickstein, M. (1967). Photic evoked potentials in the cat: evidence for a direct geniculate input to Visual II. *Exp. Neurol.*, **19**, 188–198. [111]

Berlucchi, G. see Mascetti, G. G.

Bernheimer, S. (1899). Die Wurzelgebiete der Augennerven, ihre Verbindungen und ihr Anschluss an die Gehirnrinde. *Hb. d. ges. Augenheilk.* (*Graefe-Saemisch*), **1**, Abt. **2**, Kap. **6**, 1–115. [91]

Berry, R. N. see Riggs, L. A.

Berson, E. L. see Goldstein, E. B.

Bertrand, C. see Hécaen, H.

von Bezold, W. (1873). Über das Gesetz der Farbenmischung und die physiologischen Grundfarben. *Ann. Phys., Lpz.*, **150**, 221–247.
[145]

Bidwell, S. (1897). On the negative after-images following brief retinal excitation. *Proc. Roy. Soc.*, **61**, 268–271. [149]

Biersdorf, W. R. see Armington, J. C.

Bignall, K. E., Imbert, M. & Buser, P. (1966). Optic projections to non-visual cortex of the cat. *J. Neurophysiol.*, **29**, 396–409. [111]

Bilge, M., Bingle, A., Seneviratne, K. N. & Whitteridge, D. (1967). A map of the visual cortex in the cat. *J. Physiol.*, **191**, 116P–118P.
[111]

Bingle, A. see Bilge, M.

Birke, R. see Pilz, A.

Bishop, G. H. see Clare, M. H.

Bishop, P. O., Burke, W. & Davis, R. (1962). The identification of single units in central visual pathways. *J. Physiol.*, **162**, 409–431. [104]

Bishop, P. O., Jeremy, D. & McLeod, J. G. (1953). Phenomenon of repetitive firing in lateral geniculate of cat. *J. Neurophysiol.*, **16**, 437–447. [101]

Bishop, P. O., Kozak, W., Levick, W. R. & Vakkur, G. J. (1962). The determination of the projection of the visual field on to the lateral geniculate nucleus in the cat. *J. Physiol.*, **163**, 503–539. [100]

Bishop, P. O. see also Kozak, W., Ogawa, T.

Blachowski, S. (1913). Studien über den Binnenkontrast. *Z. Sinnesphysiol.*, **47**, 291–300. [185]

Black, P. & Myers, R. E. (1964). Visual function of the forebrain commissures in the chimpanzee. *Science*, **146**, 799–800. [128]

Blake, L. (1959). The effect of lesions of the superior colliculus on brightness and pattern discriminations in the cat. *J. comp. physiol. Psychol.*, **52**, 272–278. [93]

Blakemore, C. & Campbell, F. W. (1968). Adaptation to spatial stimuli. *J. Physiol.*, **200**, 11P–13P. [169, 170]

Blakemore, C. see also Barlow, H. B.

Blakemore, C. B. & Rushton, W. A. H. (1965). The rod increment threshold during dark adaptation in normal and rod monochromat. *J. Physiol.*, **181**, 629–640. [182]

Blaurock, A. E. & Wilkins, M. H. F. (1969). Structure of frog photoreceptor membranes. *Nature*, **223**, 906–909. [23, 25]

Bliss, A. F. (1946). The chemistry of daylight vision. *J. gen. Physiol.*, **29**, 277–297. [39]

— (1948). The mechanism of retinal vitamin A formation. *J. Biol. Chem.*, **172**, 165–178. [14]

Bliss, A. F. (1949). Reversible enzymic reduction of retinene to vitamin A. *Biol. Bull.*, **97,** 221–222. [15, 39]
— (1951). Properties of the pigment layer factor in the regeneration of rhodopsin. *J. biol. Chem.*, **193,** 525–531. [18]
Blottiau, F. (1947). Les défauts d'additivité de la colorimetrie trichromatique. *Rev. Opt. (théor. instr.)*, **26,** 193–201. [216]
— see also Tessier, M.
Blum, J. S. see Chow, K. -L.
Blum, R. A. see Chow, K. -L.
Boehm, F., Sigg, B. & Monnier, M. (1944). Elektroretinographische Untersuchungen am wachen Meerschweinchen. *Helv. physiol. Acta.*, **2,** 481–493. [72]
Bogumill, G. P. see Jameson, H. D.
Boll, F. (1876). Zur Anatomie und Physiologie der Retina. *Monatsber. preuss. Akad. Wiss. Berlin*, **1876,** 783–787. [1, 33]
Bongard, M. M. & Smirnov, M. S. (1956). The four-dimensionality of human colour space (in Russian). *Dokl. Akad. Nauk S.S.S.R.*, **108,** 47–449. [204]
— (1957). Spectral sensitivity curves for the receptors connected with single fibres of the optic nerve of the frog (in Russian). *Biofizika*, **2,** 336–341. [85]
Bonnet, V. & Briot, R. (1969). Étude des composantes intracorticales tardives de la réponse de l'aire visuelle du chat a une volée d'influx afferents. *Arch. ital. Biol.*, **107,** 105–135 and 136–157. [115]
Bonting, S. L. & Bangham, A. D. (1967). On the biochemical mechanism of the visual process. *Exp. Eye Res.*, **6,** 400–413. [48, 49]
Bonting, S. L. see also Daemen, F. J. M.
Borenstein, P. see Buser, P.
Bornschein, H. (1958). Spontan-und Belichtungsaktivität in Einzelfasern des N. opticus der Katze. I. Der Einfluss kurzdauernder retinaler Ischämie. *Z. Biol.*, **110,** 210–222. [81]
Bouguer, P. (1760). *Traité d'optique sur la gradation de la lumière*. Guérin & Delatour, Paris. [175]
Bouman, M. A. (1950). Peripheral contrast thresholds of the human eye. *J. opt. Soc. Amer.*, **40,** 825–832. [169]
— (1952*a*) Peripheral contrast thresholds for various and different wavelengths for adapting field and test stimulus. *J. opt. Soc. Amer.*, **42,** 820–831. [169, 176]
— (1952*b*) Mechanisms in peripheral dark adaptation. *J. opt. Soc. Amer.*, **42,** 941–950. [185, 196]
Bouman, M. A. & van der Velden, H. A. (1947). The two-quanta explanation of the dependence of the threshold values and visual acuity on the visual angle and the time of observation. *J. opt. Soc. Amer.*, **37,** 908–919. [193, 194]
— (1948). The two-quanta hypothesis as a general explanation for the behaviour of threshold values and visual acuity for the several receptors of the human eye. *J. opt. Soc. Amer.*, **38,** 570–581. [188]

Bownds, D. (1967a). Site of attachment of retinal in rhodopsin. *Nature*, **216**, 1178–1181. [21]

— (1967b). Thesis, Harvard University, cited by Hubbard (1969) q.v. [4]

Bownds, D. see also Hubbard, R.

Boycott, B. B. & Dowling, J. E. (1969). Organization of the primate retina: light microscopy. *Philos. Trans. Roy. Soc. B.*, **255**, 109–184. [57]

Boycott, B. B. see also Dowling, J. E.

Boyle, Robert (1664). *Experiments and considerations touching colours.* Herringman, London. [200]

Boynton, R. M., Das, S. R. & Gardiner, J. (1966). Interactions between visual mechanisms revealed by mixing conditioning fields. *J. opt. Soc. Amer.*, **56**, 1775–1780. [176, 234]

Boynton, R. M., Ikeda, M. & Stiles, W. S. (1964). Interactions among chromatic mechanisms as inferred from positive and negative increment thresholds. *Vision Res.*, **4**, 87–117. [176, 231, 233, 234]

Boynton, R. M. & Kaiser, P. K. (1968). Vision: the additivity law made to work for heterochromatic photometry with bipartite fields. *Science*, **161**, 366–368. [144]

Boynton, R. M. & Riggs, L. A. (1951). The effect of stimulus area and intensity upon the human retinal response. *J. exp. Psychol.*, **42**, 217–226. [64]

Brattgård, S. -O. (1951). The importance of adequate stimulation for the chemical composition of ganglion cells. *Exptl. Cell Res.*, **2**, 693–695. [60]

Braunstein, E. P. (1923). Zur Lehre von den Kurzdauernden Lichtreizen der Netzhaut. *Z. Sinnesphysiol.*, **55**, 185–229. [169]

Bridgeman, M. see Smith, K. U.

Bridges, C. D. B. (1957). Cationic extracting agents for rhodopsin and their mode of action. *Biochem. J.*, **66**, 375–383. [3]

— (1960). Regeneration of visual pigments from their low-temperature photoproducts. *Nature*, **186**, 292–294. [23]

— (1961). Production of thermally stable photosensitive pigments in flash-irradiated solutions of frog rhodopsin. *Biochem. J.*, **79**, 135–143. [31]

— (1962a). Visual pigments of the pigeon (*Columba livia*). *Vision Res.*, **2**, 125–137. [2, 40]

— (1962b) Studies on the flash-photolysis of visual pigments 4. Dark reactions following the flash-irradiation of frog rhodopsin in suspensions of isolated photoreceptors. *Vision Res.*, **2**, 215–232. [27, 29]

— (1965). The grouping of fish visual pigments about preferred positions in the spectrum. *Vision Res.*, **5**, 223–238. [36, 37]

Bridges, C. D. B. see also Arden, G. B.

Brindley, G. S. (1952). The Bunsen-Roscoe law for the human eye at very short durations. *J. Physiol.*, **118**, 135–139. [171]

— (1953). The effects on colour vision of adaptation to very bright lights. *J. Physiol.*, **122**, 332–350. [216, 217, 218, 221, 236]

Brindley, G. S. (1954a). The summation areas of human colour-receptive mechanisms at increment threshold. *J. Physiol.*, **124**, 400–408.     [56, 241, 242, 243]

— (1954b). The order of coincidence required for visual threshold. *Proc. phys. Soc.*, **67B**, 673–676.     [190, 195]

— (1955a). The site of electrical excitation of the human eye. *J. Physiol.*, **127**, 189–200.     [53, 155, 156]

— (1955b). A photochemical reaction in the human retina. *Proc. phys. Soc.*, **68B**, 862–870.     [219, 220, 221, 251, 258]

— (1955c). The colour of light of very long wavelength. *J. Physiol.*, **130**, 35–44.     [224]

— (1956a). The passive electrical properties of the frog's retina, choroid and sclera for radial fields and currents. *J. Physiol.*, **134**, 339–352.     [66]

— (1956b). The effect on the frog's electroretinogram of varying the amount of retina illuminated. *J. Physiol.*, **134**, 353–359.     [64, 65, 75]

— (1956c). Responses to illumination recorded by micro-electrodes from the frog's retina. *J. Physiol.*, **134**, 360–384.     [50, 66, 75, 80]

— (1957a). The passive electrical properties of the retina and the sources of the electroretinogram. *Bibliotheca ophthalmologica*, **48**, 24–31.     [67]

— (1957b). Additivity in the electroretinogram. *J. Physiol.* **137**, 51–52P.     [64, 65]

— (1957c). Two theorems in colour vision. *Quart. J. exp. Psychol.*, **9**, 101–104.     [210, 258]

— (1958). The sources of slow electrical activity in the frog's retina. *J. Physiol.*, **140**, 247–261.     [66]

— (1959a). The discrimination of after-images. *J. Physiol.*, **147**, 194–203.     [31, 150, 172]

— (1959b). Die Physiologie des Farbensehens. *Klin. Wschr.*, **37**, 1157–1162.     [212]

— (1960a). Two more visual theorems. *Quart. J. exp. Psychol.*, **12**, 110–112.     [226]

— (1960b). Observations first reported in the first edition of this book.     [27, 29, 155, 214

— (1961). On- and off-effects in stimulation of the eye by alternating current. *J. Physiol.*, **158**, 3P–5P.     [174]

— (1962a). Beats produced by simultaneous stimulation of the human eye with intermittent light and intermittent or alternating electric current. *J. Physiol.*, **164**, 157–167.     [173, 245]

— (1962b). Two new properties of foveal after-images and a photo-chemical hypothesis to explain them. *J. Physiol.*, **164**, 168–179.     [151]

— (1963a). Red-sensitive retinal pigments in the frog, Jersey bull calf and East African baboon. *J. Physiol.*, **167**, 49P–50P.     [43, 44, 254]

— (1963b). The relation of frequency of detection to intensity of stimulus

for a system of many detectors each of which is stimulated by a
m-quantum coincidence. *J. Physiol.*, **169**, 412–415.                    [195]

Brindley, G. S. (1964). A new interaction of light and electricity in
stimulating the human retina. *J. Physiol.*, **171**, 514–520.          [157]

— (1966). The deformation phosphene and the funnelling of light into
rods and cones. *J. Physiol.*, **188**, 24P–25P.                    [160]

Brindley, G. S., Carpenter, R. H. S. & Rushton, D. N. (1967). Reaction
times for simple shape discriminations requiring one or both
visual cortices. *Q. J. exp. Psychol.*, **19**, 70–72.              [127]

Brindley, G. S., Du Croz, J. & Rushton, W. A. H. (1966). The flicker
fusion frequency of the blue-sensitive mechanism of colour vision.
*J. Physiol.*, **183**, 497–500.                           [173, 244]

Brindley, G. S. & Gardner-Medwin, A. R. (1965). Evidence that the
'early receptor potential' of the retina does not depend on changes
in the permeability of membranes to specific ions. *J. Physiol.*, **180**,
1P.                                      [31]

— (1966). The origin of the early receptor potential. *J. Physiol.*, **182**,
185–194.                                    [31, 32]

Brindley, G. S., Gautier-Smith, P. C. & Lewin, W. (1969). Cortical
blindness and the functions of the non-geniculate fibres of the
optic tracts. *J. Neurol. Neurosurg. Psychiatr.*, **32**, 259–264.   [91, 93]

Brindley, G. S. & Hamasaki, D. I. (1962). Evidence that the cat's electroretinogram is not influenced by impulses passing to the eye along
the optic nerve. *J. Physiol.*, **163**, 558–565.                    [109]

— (1963). The properties and nature of the R membrane of the frog's
eye. *J. Physiol.*, **167**, 599–606.                            [66]

— (1966). Histological evidence against the view that the cat's optic
nerve contains centrifugal fibres. *J. Physiol.*, **184**, 444.        [107]

Brindley, G. S. & Lewin, W. S. (1968). The sensations produced by
electrical stimulation of the visual cortex. *J. Physiol.*, **196**, 479–493.
                                 [99, 110, 114, 124, 126, 149]

Brindley, G. S. & Rushton, W. A. H. (1959). The colour of monochromatic light when passed into the human retina from behind. *J. Physiol.*,
**147**, 204–208.                                 [250, 253]

Brindley, G. S. & Westheimer, G. (1965). The spatial properties of the
human electroretinogram. *J. Physiol.*, **179**, 518–537.            [64]

Brindley, G. S. & Willmer, E. N. (1952). The reflexion of light from the
macular and peripheral fundus oculi in man. *J. Physiol.*, **116**,
350–356.                                     [28, 140]

Brindley, G. S. see also Barlow, H. B., Rushton, W. A. H., Wolff, J. G.

Briot, R. see Bonnet, V.

Broca, A. & Sulzer, D. (1902). La sensation lumineuse en fonction du
temps. *J. Physiol. Path. gén.*, **4**, 632–640.                    [143]

Broda, E. E. & Goodeve, C. F. (1941). The behaviour of visual purple at
low temperature. *Proc. Roy. Soc.*, **179A**, 151–159.                [9]

Broda, E. E., Goodeve, C. F. & Lythgoe, R. J. (1940). The weight of the
chromophore carrier in the visual purple molecule. *J. Physiol.*,
**98**, 397–404.                                   [3, 4]

Brodskiǐ, V. Ya. & Nechaeva, N. V. (1958). Quantitative cytochemical investigation of ribonucleic acid in ganglion cells of the retina during normal functioning and fatigue (in Russian). *Biofizika*, **3,** 269–273.                                                                              [60]

Brodski, V. Ya. see also Utina, I. A.

Brooke, R. N. L., Downer, D. J. de C. & Powell, T. P. S. (1965). Centrifugal fibres to the retina in the monkey and cat. *Nature*, **207,** 1365–1367.                                                                              [106]

Brouwer, B. & Zeeman, W. P. C. (1926). The projection of the retina in the primary optic neuron in monkeys. *Brain*, **49,** 1–35. [99, 100]

Brown, J. L., Kuhns, M. P. & Adler, H. E. (1957). Relation of threshold criterion to the functional receptors of the eye. *J. opt. Soc. Amer.*, **47,** 198–204.                                                                              [167]

Brown, J. E. & Major, D. (1966). Cat retinal ganglion cell dendritic fields. *Exp. neurol.*, **15,** 70–78.                                                [57]

Brown, J. E. see also Dowling J. E.

Brown, K. T. & Murakami, M. (1964). A new receptor potential of the monkey retina with no detectable latency. *Nature*, **201,** 626–628.
                                                                              [31]

Brown, K. T. & Watanabe, K. (1962). Isolation and identification of a receptor potential from the pure cone fovea of the monkey retina. *Nature*, **193,** 958–960.                                                  [69]

Brown, K. T. & Wiesel, T. N. (1958). Intraretinal recording in the unopened cat eye. *Amer. J. Ophthal.*, **46,** 91–96.           [72, 76]

— (1959). Intraretinal recording with micropipette electrodes in the intact cat eye. *J. Physiol.*, **149,** 537–562.                    [66, 75, 80]

— (1961). Analysis of the intraretinal electroretinogram in the intact cat eye. *J. Physiol.*, **158,** 229–256.                              [66, 68, 72]

Brown, K. T. see also Ogden, T. E.

Brown, P. K., Gibbons, R. I. & Wald, G. (1963). The visual cells and visual pigments of the mudpuppy, Necturus. *J. cell. Biol.*, **19,** 79–106.
                                                                              [24]

Brown, P. K. & Wald, G. (1963). Visual pigments in human and monkey retinas. *Nature*, **200,** 37–43.                                        [43]

— (1964). Visual pigments in single rods and cones of the human retina. *Science*, **144,** 45–52.                      [39, 44, 221, 254]

Brown, P. K. see also Cone, R. A., Matthews, R. G., Wald, G.

Brown, R. H. see Graham, C. H.

Brown, W. R. J. (1957). Color discrimination of twelve observers. *J. opt. Soc. Amer.*, **47,** 137–143.                                        [231]

Brown, W. R. J. & MacAdam, D. L. (1949). Visual sensitivities to combined chromaticity and luminance differences. *J. opt. Soc. Amer.*, **39,** 808–834.                                                                              [231]

Brücke, E. W. (later von Brücke), (1864). Über den Nutzeffekt intermittierender Netzhautreizungen. *S. B. Akad. Wiss. Wien, Math. Naturw. Kl.*, **49,** Part **2,** 128–153.                                        [143]

von Brücke, E. T. & Garten, S. (1907). Zur vergleichenden Physiologie der Netzhautströme. *Pflügers Arch. ges. Physiol.*, **120**, 290–348. [61]

von Brücke, E. T. see also Cords, R.

Bucy, P. C. see Halstead, W. C., Klüver, H.

Bulanova, K. N. & Luizov, A. V. (1954). The liminal duration of darkening of a point source of light (in Russian). *Dokl. Akad. Nauk S.S.S.R.*, **98**, 205–206.                                                                [176]

Burch, G. (1898). On artificial temporary colour-blindness, with an examination of the colour sensations of 109 persons. *Phil. Trans.*, **191B**, 1–34.                                                                          [236]

Burch, J. M. see Stiles, W. S.

Burke, W. & Hayhow, W. R. (1968). Disuse in the lateral geniculate nucleus of the cat. *J. Physiol.*, **194**, 495–519.                           [81]

Burke, W. & Sefton, Ann J. (1966a). Discharge patterns of principal cells and interneurones in lateral geniculate nucleus of rat. *J. Physiol.*, **187**, 201–212.                                                         [102]

— (1966b). Recovery of responsiveness of cells of lateral geniculate nucleus of rat. *J. Physiol.*, **187**, 213–229.                           [102]

— (1966c). Inhibitory mechanisms in lateral geniculate nucleus of rat. *J. Physiol.*, **187**, 231–246                                              [102]

Burleigh, S. see Forbes, A.

Burns, B. D., Heron, W. & Pritchard, R. (1962). Physiological excitation of visual cortex in cat's unanaesthetized isolated forebrain. *J. Neurophysiol.*, **25**, 165–181.                                                    [120]

Butler, C. R. (1968). A memory-record for visual discrimination habits produced in both cerebral hemispheres of monkey when only one hemisphere has received direct visual information. *Brain Res.*, **10**, 152–167.                                                                [128]

Byzov, A. L. (1959). Analysis of the distribution of potentials and currents within the retina during stimulation by light. I. The activity of bipolar cells of two types (in Russian). *Biofizika*, **4**, 689–701.     [61]

— (1968). Localization of the R membrane in the frog eye by means of an electrode marking method. *Vision Res.*, **8**, 697–700.           [66]

Byzov, A. L. & Hanitzsch, R. (1964). Die elektrischen Eigenschaften und die Struktur der R-Membran. *Vision Res.*, **4**, 483–492.           [66]

Byzov, A. L. & Trifonov, Yu. A. (1968). The response to electric stimulation of horizontal cells in the carp retina. *Vision Res.*, **8**, 817–822.
                                                                        [79]

Byzov, A. L., see also Utina, I. A.

Cajal, S. Ramón (1889). Sur la morphologie et les connexions des elements de la retine des oiseaux. *Anat. Anzeiger*, **4**, 111–121.
                                                                        [105]

— (1894). *Die Retina der Wirbelthiere*. Bergmann, Wiesbaden.      [57]

— (1904). *Textura del sistema nervioso del hombre y de los vertebrados*. Moya, Madrid.                                                          [98]

Calvet, J., Cathala, H. P., Hirsch, J. & Scherrer, J. (1956). La réponse corticale visuelle de l'homme etudiée par une methode d'integration. *C. R. Soc. Biol.*, **150**, 1348–1351.                                   [116]

Campbell, F. W. & Green, D. G. (1965). Optical and retinal factors affecting visual resolution. *J. Physiol.*, **181**, 576–593. [168]

Campbell, F. W. & Gubisch, R. W. (1967). The effect of chromatic aberration on visual acuity. *J. Physiol.*, **192**, 345–358. [167]

Campbell, F. W., Kulikowski, J. J. & Levinson, J. (1966). The effect of orientation on the visual resolution of gratings. *J. Physiol.*, **187**, 427–436. [168]

Campbell, F. W. & Rushton, W. A. H. (1955). Measurement of the scotopic pigment in the living human eye. *J. Physiol.*, **130**, 131–147. [180]

Campbell, F. W. see also Alpern, M., Blakemore, C., Rushton, W. A. H.

von Campenhausen, C. (1969). The colors of Benham's top under metameric illuminations. *Vision Res.*, **9**, 677–682. [155]

Carpenter, R. H. S. see Brindley, G. S., Wolff, J. G.

Cathala, H. P. see Calvet, J.

Cavaggioni, A. (1968). The dark-discharge of the eye in the unrestrained cat. *Pflügers Arch. ges. Physiol.*, **304**, 75–80. [108]

Chase, A. M. & Smith, E. L. (1939). Regeneration of visual purple in solution. *J. gen. Physiol.*, **23**, 21–39. [17]

Chase, A. M. see also Shlaer, S.

Choudhury, B. P., Whitteridge, D. & Wilson, M. E. (1965). The functions of the callosal connections of the visual cortex. *Q. J. exp. Physiol.*, **50**, 214–219. [127]

Clare, M. H. & Bishop, G. H. (1954). Responses from an association area secondarily activated from optic cortex. *J. Neurophysiol.*, **17**, 271–277. [111]

Clark, W. E. Le Gros, McKeown, T. & Zuckerman, S. (1939). Visual pathways concerned in gonadal stimulation in ferrets. *Proc. Roy. Soc. B.*, **126**, 449–468. [95]

Clark, W. E. Le Gros & Penman, G. G. (1934). The projection of the retina in the lateral geniculate body. *Proc. Roy. Soc.*, **114B**, 291–313. [99, 100]

Clarke, F. J. J. (1963). Further studies of extra-foveal colour metrics. *Optica Acta*, **10**, 257–274. [204, 216]

Cobb, P. W. (1915). The influence of pupillary diameter on visual acuity. *Amer. J. Physiol.*, **36**, 335–346. [167]

Cobb, W. A. Ettlinger, G. & Morton, H. B. (1967). Cerebral potentials evoked by pattern reversal and their suppression in visual rivalry. *Nature*, **216**, 1123–1125. [116]

Cobbs, W. H. see Cone, R. A.

Cohen, A. J. (1964). Some observations on the fine structure of the retinal receptors of the American gray squirrel. *Invest. Ophthal.*, **3**, 198–216. [40]

— (1965). A possible cytological basis for the R membrane in the vertebrate eye. *Nature*, **205**, 1222–1223. [66]

— (1968). New evidence supporting the linkage to extra-cellular space of outer segment saccules of frog cones but not rods. *J. cell. Biol.*, **37**, 424–444. [25, 33]

Cohen, B., Feldman, M. & Diamond, S. P. (1969). Effects of eye movement, brain-stem stimulation and alertness on transmission through lateral geniculate body of monkey. *J. Neurophysiol.*, **32**, 583–594.
[103]

Cohen, J. (1946). Color adaptation to stimuli of different spectral composition but equal I.C.I. specification. *J. opt. Soc. Amer.*, **36**, 717.
[214]

Cole, B. L. see Henry, G. H.

Cole, M. Schutta, H. S. & Warrington, E. K. (1962). Visual disorientation in homonymous half-fields. *Neurology*, **12**, 257–263. [131]

Colman, A. D. see Hubbard, R.

Colonnier, M. & Guillery, R. W. (1964). Synaptic organization in the lateral geniculate nucleus of the monkey. *Z. Zellforsch.*, **62**, 333–355.
[98]

Collins, F. D. (1953). Rhodopsin and indicator yellow. *Nature*, **171**, 469–471. [19]

Collins, F. D., Love, R. M. & Morton, R. A. (1952). Studies in rhodopsin. 4: Preparation of rhodopsin. *Biochem. J.*, **51**, 292–298. [4]

Collins, F. D. & Morton, R. A. (1950*a*). Studies on rhodopsin. 2: Indicator yellow. *Biochem., J.*, **47**, 10–17. [19]

— (1950*b*). Studies on rhodopsin. 3: Rhodopsin and transient orange. *Biochem. J.*, **47**, 18–24. [9, 23]

Collins, F. D. see also Ball, S.

Commichau, R. (1955). Adaptationszustand und Unterschiedsschwellenenergie für Lichtblitze. *Z. Biol.*, **108**, 145–160. [176]

Cone, R. A. (1963). Quantum relations of the rat electroretinogram. *J. gen. Physiol.*, **46**, 1267–1286. [54]

— (1964). Early receptor potential of the vertebrate retina. *Nature*, **204**, 736–739. [31]

— (1967). Early receptor potential: photoreversible charge displacement in rhodopsin. *Science*, **155**, 1128–1131. [32]

Cone, R. A. & Brown, P. K. (1967). Dependence of the early receptor potential on the orientation of rhodopsin. *Science*, **156**, 536. [32]

— (1969). Spontaneous regeneration of rhodopsin in the isolated retina. *Nature*, **221**, 818–820. [27, 34]

Cone, R. A. & Cobbs, W. H. (1969). Rhodopsin cycle in the living eye of the rat. *Nature*, **221**, 820–822. [27]

Cone, R. A. & Ebrey, T. G. (1965). Functional independence of the two major components of the rod electroretinogram. *Nature*, **206**, 913–915. [54]

Cone, R. A. & Platt, J. R. (1964). Rat electroretinogram: Evidence for separate processes governing b-wave latency and amplitude. [75]

Conner, J. P. & Ganoung, R. E. (1935). An experimental determination of the visual thresholds at low values of illumination. *J. opt. Soc. Amer.*, **25**, 287–294. [166, 168]

Cords, R. & von Brücke, E. T. (1907). Über die Geschwindigkeit des Bewegungsnachbildes. *Pflügers Arch. ges. Physiol.*, **119**, 54–76.
[146]

Cornsweet, J. C. see Riggs, L. A.

Cornsweet, T. N., Fowler, H., Rabedeau, R. G., Whalen, R. E. & Williams, D. R. (1958). Changes in the perceived color of very bright stimuli. *Science*, **128**, 898–899. [145]

Cornsweet, T. N. see also Riggs, L. A. and Fig. 6.3.

Cowan, W. M. & Powell, T. P. S. (1963). Centrifugal fibres in the avian visual system. *Proc. Roy. Soc.*, **158B**, 232–252. [106]

Cowan, W. M. see also Dowling, J. E.

Cowey, A. & Weiskrantz, L. (1967). A comparison of the effects of inferotemporal and striate cortex lesions on the visual behaviour of rhesus monkeys. *Q. J. exp. Psychol.*, **19**, 246–253. [129]

Craik, K. & Vernon, M. (1941). The nature of dark adaptation. *Brit. J. Psychol.*, **32**, 62–81. [178]

Crawford, B. H. (1937). The change of visual sensitivity with time. *Proc. Roy. Soc.*, **123B**, 69–89. [185]

— (1940). Ocular interaction in its relation to measurements of brightness threshold. *Proc. Roy. Soc.*, **128B**, 552–559. [109]

— (1946). Photochemical laws and visual phenomena. *Proc. Roy. Soc.*, **133B**, 63–75. [180, 184]

— (1947). Visual adaptation in relation to brief conditioning stimuli. *Proc. Roy. Soc.*, **134B**, 283–302. [185]

— see also Stiles, W. S.

Crescitelli, F. (1967). Extraction of visual pigments with certain alkyl phenoxy polyethoxy ethanol surface-active compounds. *Vision Res.*, **7**, 685–693. [3]

Crescitelli, F., Mommaerts, W. F. H. M. & Shaw, T. I. (1966). Circular dichroism of visual pigments in the visible and ultraviolet spectral regions. *Proc. Nat. Acad. Sci.*, **56**, 1729–1734. [5]

Crescitelli, F. & Shaw, T. I. (1964). The circular dichroism of some visual pigments. *J. Physiol.*, **175**, 43P–45P. [5]

van Crevel, H. (1958). *The rate of secondary degeneration in the central nervous system.* Leiden: E. Ijdo. [106]

Critchlow, V. (1963). The role of light in the neuroendocrine system. *Advances in Neuroendocrinology*. Urbana: Univ. of Illinois Press, pp 377–402. [95]

Cruz, A. see Moreland, J. D.

Curtis, H. J. (1940). Intercortical connections of corpus callosum as indicated by evoked potentials. *J. Neurophysiol.*, **3**, 407–413. [127]

Daemen, F. J. M. & Bonting, S. L. (1968). Specificity of the retinaldehyde effect on cation movements in rod outer segments. *Biochim. Biophys. Acta.*, **163**, 212–217. [48]

Dalvi, D. P. see Ball, S.

Dartnall, H. J. A. (1952). Visual pigment 467, a photosensitive pigment present in tench retinae. *J. Physiol.*, **116**, 257–289. [37, 38]

— (1960). Visual pigment from a pure-cone retina. *Nature*, **188**, 475–479. [40]

— (1961). Visual pigments before and after extraction from visual cells. *Proc. Roy. Soc.*, **154B**, 250–266. [26]

Dartnall, H. J. A. & Lythgoe, J. N. (1965). The spectral clustering of visual pigments. *Vision Res.*, **5**, 81–100.                          [36, 37]

Das, S. R. see Boynton, R. M.

David, H. see Alpern, M.

Davis, R. see Bishop, P. O.

Daw, N. W. (1967). Goldfish retina: organization for simultaneous contrast. *Science*, **158**, 942–944.                          [85]

Delacour, J. see Wolff, J. G.

Dell, P. see Dumont, S.

Denney, D., Baumgartner, G. & Adorjani, C. (1968). Responses of cortical neurones to stimulation of the visual afferent radiations. *Exp. Brain Res.*, **6**, 265–272.                          [119]

Denny-Brown, D. (1962). The mid-brain and motor integration. *Proc. Roy. Soc. Med.*, **55**, 527–538.                          [93]

Denny-Brown, D. see also Meyer, J. S.

Denton, E. J. (1954). 1. A method of easily observing the dichroism of the visual rods. 2. On the orientation of molecules in the visual rods of Salamandra maculosa. *J. Physiol.*, **124**, 16P and 17P.                          [23, 28]

— (1955*a*). Une nouvelle méthode pour determiner la courbe d'absorption du pourpre retinien. *C. R. Acad. Sci.*, **239**, 1315–1316.                          [28]

— (1955*b*). On the bleaching of the photosensitive substance in the retinal rods of Xenopus laevis. *J. Physiol.*, **131**, 6P–7P.                          [28]

Denton, E. J. & Pirenne, M. H. (1951). Spatial summation at the absolute threshold of peripheral vision. *J. Physiol.*, **116**, 32P–33P.                          [193]

Denton, E. J. & Walker, M. A. (1958). The visual pigment of the conger eel. *Proc. Roy. Soc.*, **148B**, 257–269.                          [27]

Denton, E. J. & Warren, F. J. (1957). The photosensitive pigments in the retinae of deep-sea fish. *J. mar. biol. Ass. U.K.*, **36**, 651–662.                          [36]

Denton, E. J. & Wyllie, J. H. (1955). Study of the photosensitive pigments in the pink and green rods of the frog. *J. Physiol.*, **127**, 81–89.                          [27]

De Renzi, E. & Faglioni, P. (1967). The relationship between visuospatial impairment and constructional apraxia. *Cortex*, **3**, 327–343.                          [131]

De Robertis, E. (1956). Electron microscope observations on the submicroscopic organization of the retinal rods. *J. biophys. biochem. Cytol.*, **2**, 319–330.                          [23]

Dettmar, P. see Ostrovskiĭ, M. A.

De Valois, R. (1965). Analysis and coding of color vision in the primate visual system. *Cold Spr. Harb. Symp.*, **30**, 567–579.                          [103, 104]

Diamond, S. P. see Cohen, B.

Dieterici, C. see König, A.

Ditchburn, R. W. & Ginsborg, B. L. (1952). Vision with a stabilized retinal image. *Nature*, **170**, 36–37.                          [148]

Dobelle, W. H. see Marks, W. B.

Dodt, E. (1951). Über die sekundäre Erhebung im Aktionspotential des menschlichen Auges bei Belichtung. *Graefes Arch.*, **151**, 672–692.                          [71]

Dodt, E. (1956). Geschwindigkeit der Nervenleitung innerhalb der Netzhaut. *Experientia, Basel*, **12**, 34–35.                    [107]
—  (1957). Ein Doppelinterferenzfilter-Monochromator besonders hoher Leuchtdichte. *Bibliotheca ophthal.*, **48**, 32–37.          [72]
Donders, F. C. (1881). Ueber Farbensysteme. *Graefes Arch. Ophthal.*, **27**, 155–223.                            [207, 209]
Donley, N. J. see Guth, S. L.
Donner, K. O. (1949). Variations due to colour in the spike frequency time curves of single retinal elements. *Experientia, Basel*, **5**, 413–414.                            [84]
—  (1953). The spectral sensitivity of the pigeon's retinal elements. *J. Physiol.*, **122**, 524–527.                      [83]
—  (1959). The effect of a coloured adapting field on the spectral sensitivity of frog retinal elements. *J. Physiol.*, **149**, 318–326.    [85]
Donner, K. O. & Rushton, W. A. H. (1959*a*). Retinal stimulation by light substitution. *J. Physiol.*, **149**, 288–302.            [85, 88]
—  (1959*b*). Rod-cone interaction in the frog's retina analysed by the Stiles-Crawford effect and by dark adaptation. *J. Physiol.*, **149**, 303–317.                            [85]
Donovan, A. (1967). The nerve fibre composition of the cat optic nerve. *J. Anat.*, **101**, 1–11.                        [107]
Dowling, J. E. (1960). Chemistry of visual adaptation in the rat. *Nature*, **188**, 114–118.                          [35, 180]
—  (1965). Foveal receptors of the monkey retina: fine structure. *Science*, **147**, 57–59.                            [23]
Dowling, J. E. & Boycott, B. B. (1966). Organization of the primate retina: electron microscopy. *Proc. Roy. Soc. B.*, **166**, 80–111.    [58]
Dowling, J. E., Brown, J. E. & Major, D. (1966). Synapses of horizontal cells in rabbit and cat retinas. *Science*, **153**, 1639–1641.    [59]
Dowling, J. E. & Cowan, W. M. (1966). An electron microscope study of normal and degenerating centrifugal fiber terminals in the pigeon retina. *Z. f. Zellforschung*, **71**, 14–28.                [106]
Dowling, J. E. see also Boycott, B. B., Frank, R. N., Hubbard, R., Weinstein, G. W., Werblin, F. S.
Downer, D. J. de C. see Brooke, R. N. L.
Dresler, A. (1937). *Licht*, **7**, 203. Cited by Dresler, A. (1953). The non-additivity of heterochromatic brightness. *Trans. ill. eng. Soc.*, **18**, 141–156.                          [144]
Du Bois-Reymond, E. (1848). *Untersuchungen über thierische Elektricität*, Bd. **1**, Reimer, Berlin.                      [61]
Du Croz, J. J. see Brindley, G. S.
Dumont, S. & Dell, P. (1958). Facilitations spécifiques et non-spécifiques des réponses visuelles corticales. *J. Physiol.* (*Paris*), **50**, 261–264.
                                [113]
Dunn, R. F. (1969). The dimensions of rod outer segments related to light absorption in the gecko retina. *Vision Res.*, **9**, 603–609.  [221]
Durell, J. see Wald, G.

Easter, S. S. (1968). Excitation in the goldfish retina: evidence for a non-linear intensity code. *J. Physiol.*, **195**, 253–271. [164]

Ebe, M. see Motokawa, K.

Ebrey, T. G. (1968). The thermal decay of the intermediates of rhodopsin *in situ. Vision Res.*, **8**, 965–982. [32]

Ebrey, T. G. see also Cone, R. A., Pak, W. L.

Einthoven, W. & Jolly, W. A. (1908). The form and magnitude of the electrical response of the eye to stimulation by light at various intensities. *Quart. J. exp. Physiol.*, **1**, 373–416. [61, 62, 71]

Emsley, H. H. (1925). Irregular astigmatism of the eye: effect of correcting lenses. *Trans. opt. Soc. (Lond.)*, **27**, 28–41. [168]

Engelking, E. (1926). Die Tritanomalie, ein bisher unbekannter Typus der anomalen Trichromasie. *Graefes Arch. Ophthal.*, **116**, 196. [229]

Enoch, J. M. (1966). Validation of an indicator of mammalian retinal receptor response: density of stain as a function of stimulus magnitude. *J. opt. Soc. Amer.*, **56**, 116–123. [61]

Enoch, J. M., Goldmann, H. & Sunga, R. (1969). The ability to distinguish which eye was stimulated by light. *Invest. Ophthal.*, **8**, 317–331. [152, 248]

Enoch, J. M. & Stiles, W. S. (1961). The colour change of monochromatic light with retinal angle of incidence. *Optica Acta*, **8**, 329–358. [221, 248]

Enroth, C. (1952). The mechanism of flicker and fusion studied on single retinal elements in the dark-adapted eye of the cat. *Acta. physiol. Scand.*, **27**, Suppl. **100**, 1–67. [88]

Erhardt, F. see Ostroy, S. E.

Ernst, W. (1968). The dependence of critical flicker frequency and the rod threshold on the state of adaptation of the eye. *Vision Res.*, **8**, 889–900. [182]

Ernst, W. see also Arden, G. B.

Etingov, R. N., Shukolyukov, S. A. & Leontyev, V. G. (1964). Efflux of sodium and potassium ions from outer segments of photoreceptors of the retina under the influence of illumination and vitamin A (in Russian). *Doklady Akad. Nauk. S.S.S.R.*, **156**, 979–981. [48]

Ettlinger, G. (1959). Visual discrimination following successive temporal ablations in monkeys. *Brain*, **82**, 232–250. [130]

Ettlinger, G., Iwai, E., Mishkin, M. & Rosvold, H. E. (1968). Visual discrimination in the monkey following serial ablation of inferotemporal and preoccipital cortex. *J. comp. physiol. Psychol.*, **65**, 110–117. [130]

Ettlinger, G. & Kalsbeck, J. E. (1962). Changes in tactile discrimination and in visual reaching after successive and simultaneous bilateral posterior parietal ablations in the monkey. *J. Neurol. Neurosurg. Psychiatr.*, **25**, 256–268. [131]

Ettlinger, G. see also Bates, J. A. V., Cobb, W. A.

Evarts, E. V. see Schoolman. A.

Ewald, A. & Kühne, W. (1878a, b, c). Untersuchungen über den Sehpurpur. I, II, III, *Untersuchungen physiol. Inst. Univ. Heidelberg*, **1**, 139–218, 248–290, 370–455.        [1, 17, 33]

Exner, F. (1902). Über die Grundempfindungen im Young-Helmholtz'-chen Farbensystem. *S. B. Akad. Wiss. Wien.*, **111**, Abt. 2a, 857–877.        [145]

Fabian, M. see Stone, J.

Faglioni, P. see De Renzi, E.

Falk, G. & Fatt, P. (1966). Rapid hydrogen ion uptake of rod outer segments and rhodopsin solutions on illumination. *J. Physiol.*, **183**, 211–224.        [33]

— (1968a). Passive electrical properties of rod outer segments. *J. Physiol.*, **198**, 627–646.        [49]

— (1968b). Conductance changes produced by light in rod outer segments. *J. Physiol.*, **198**, 647–699.        [49]

Falls, H. F., Wolter, J. R. & Alpern, M. (1965). Typical total monochromacy. *Arch. Ophthal.*, **74**, 610–616.        [227]

Fang, H. C. see Meyer, J. S.

Fatt, P. see Falk, G.

Fechner, G. T. (1838). Über eine Scheibe zur Erzeugung subjectiver Farben. *Ann. Phys. Lpz.*, **45**, 227–232.        [154]

Fedorov, N. T., Sklyarevich, V. V., Yuryev, M. A. & Mashirova, O. F. (1953). Basic regularities in the phenomena of colour contrast (in Russian). *Probl. fiziol. Opt.*, **8**, 75–91.        [152]

Feldberg, V. & Vogt, M. (1948). Acetycholine synthesis in different regions of the central nervous system. *J. Physiol.*, **107**, 372–381. [97]

Feldman, M. see Cohen, B.

Festing, E. R. see Abney, W. de W.

Fiorentini, A., Jeanne, M. & Toraldo di Francia, G. (1955). Mesures photometriques visuelles sur un champ à gradient d'éclairement variable. *Optic. Acta*, **1**, 192–193.        [185]

Fischer, M. H. (1932). Elekrobiologische Erscheinungen an der Hirnrinde. I. *Pflügers Arch. ges. Physiol.*, **230**, 161–178.        [115, 116]

Fitzhugh, R. see Barlow, H. B., Kuffler, S. W.

Flamant, F. & Stiles, W. S. (1948). The directional and spectral sensitivities of the retinal rods to adapting fields of different wavelengths. *J. Physiol.*, **107**, 187–202.        [249]

Foerster, O. (1929). Beiträge zur Pathophysiologie der Sehbahn und der Sehsphäre. *J. Psychol. Neurol. Lpz.*, **39**, 463–485.        [110, 124]

Foerster, O. & Penfield, W. (1930). Der Narbenzug am und im Gehirn bei traumatischer Epilepsie in seiner Bedeutung fur das Zustandekommen der Anfälle und für die therapeutische Bekämpfung derselben. *Z. ges. Neurol. Psychiat.*, **125**, 475–572.        [124]

da Fonseca, J. S. see Jung, R.

Forbes, L. M. see Mote, F. A.

Fowler, H. see Cornsweet, T. N.

Franchi, C. M. see de Robertis, E.

François, J. & Verriest, G. (1957). Les dyschromatopsies acquises. *Ann. oculist.*, **190**, 713–746, 812–859 and 893–943.        [230]

Frank, R. N. & Dowling, J. E. (1968). Rhodopsin photoproducts: effects on electroretinogram sensitivity in isolated perfused rat retina. *Science*, **161**, 487–489. [27, 29]

Fridrikh, L. (1957). Colour-mixture curves for normal trichromats determined by direct energy measurements (in Russian). *Biofizika*, **2**, 124–128. [203]

Fujino, T. & Hamasaki, D. I. (1965). The effect of occluding the retinal and choroidal circulations on the electroretinogram of monkeys. *J. Physiol.*, **180**, 837–845. [69, 72]

Fujita, H. (1911). Pigmentbewegung und Zapfenkontraction im Dunkelauge des Frosches bei Einwirkung verschiedener Reize. *Arch. v. vergl. Ophthal.*, **2**, 164–179. [107]

Fujii, T. see Ikeda, M.

Fulton, A. B. see Baker, H. D.

Fulton, J. F. (1928). Observations upon the vascularity of the human occipital lobe during visual activity. *Brain*, **51**, 310–320. [123]

Furukawa, T. & Hanawa, I. (1955). Effects of some common cations on electroretinogram of the toad. *Jap. J. Physiol.*, **5**, 280–300. [49]

Fusillo, M. see Geschwind, N.

Ganoung, R. E. see Conner, J. P.

Gardiner, J. see Boynton, R. M.

Gardner-Medwin, A. R. see Brindley, G. S.

Garey, L. J., Jones, E. G. & Powell, T. P. S. (1968). Interrelationships of striate and extrastriate cortex with the primary relay sites of the visual pathway. *J. Neurol. Neurosurg. Psy.*, **31**, 135–157. [95]

Garten, S. see von Brücke, E. T.

Gautier-Smith, P. C. see Brindley, G. S.

Geschwind, N. & Fusillo, M. (1966). Color-naming defects in association with alexia. *Arch. Neurol.*, **15**, 137–146. [128]

Gestring, G. F. see Jacobson, J. H.

Gibbons, R. I. see Brown, P. K.

Gibson, I. M. (1962). Visual mechanisms in a cone-monochromat. *J. Physiol.*, **161**, 10P–11P. [229]

Gibson, J. J. (1933). Adaptation, after effect and contrast in the perception of curved lines. *J. exp. Psychol.*, **16**, 1–31. [146]

Gilbert, M. (1950). Colour perception in parafoveal vision. *Proc. phys. Soc.*, **63B**, 83–89. [204]

Gildemeister, M. (1914). Über die Wahrnehmbarkeit von Lichtlücken. *Z. Sinnesphysiol.*, **48**, 256–267. [176]

Gilinsky, A. S. (1968). Orientation-specific effects of patterns of adapting light on visual acuity. *J. opt. Soc. Amer.*, **58**, 13–18. [169]

Glees, P., Hallermann, W. & Naeve, H. (1964). Die Repräsentation retinaler Sektoren im Corpus geniculatum laterale des Affen. *v. Graefes Arch. Ophthal.*, **167**, 367–376. [100]

Glezer, V. D. (1959). Cone adaptation as a nervous process (in Russian). *Dokl. Akad. Nauk. S.S.S.R.*, **126**, 1110–1113. [176]

— (1965). The receptive fields of the retina. *Vision Res.*, **5**, 497–525. [164]

Glickstein, M. see Berkley, M.

Ginsborg, B. L. see Ditchburn, R. W.

Godfraind, J. M. & Meulders, M. (1969). Effets de la stimulation sensorielle somatique sur les champs visuels des neurones de la région genouillée chez le chat anesthésié au chloralose. *Exp. Brain Res.*, **9**, 183–200.                                                        [104]

Goldby, F. (1957). A note on transneuronal atrophy in the human lateral geniculate body. *J. Neurol. Neurosurg. Psychiat.*, **20**, 202–207.    [98]

Goldmann, H. see Enoch, J.

Goldstein, E. B. (1967). Early receptor potential of the isolated frog (Rana pipiens) retina. *Vision Res.*, **7**, 837–845.                    [32]

Goldstein, E. B. (1968). Visual pigments and the early receptor potential of the isolated frog retina. *Vision Res.*, **8**, 953–963.           [33]

Goldstein, E. B. & Berson, E. L. (1969). Cone dominance of the human early receptor potential. *Nature*, **222**, 1272–1273.                 [33]

Goodeve, C. F., Lythgoe, R. J. & Schneider, E. E. (1941). The photosensitivity of visual purple solutions and the scotopic sensitivity of the eye in the ultraviolet. *Proc. Roy. Soc.*, **130B**, 380–395.         [13]

Goodeve, C. F. see also Broda, E. E., Schneider, E.

Gotch, F. (1903). The time relations of the photo-electric changes in the eyeball of the frog. *J. Physiol.*, **29**, 388–410.                    [61]

Goto, M. & Toida, N. (1954). Some mechanisms of color reception found by analyzing the electroretinogram of the frog. Part I: Colour characteristics of the multiple off-response and number of retinal elements. *Jap. P. Physiol.*, **4**, 221–228.                          [73]

Gouras, P. (1960). Graded potentials of bream retina. *J. Physiol.*, **152**, 487–505.                                                          [78]

— (1967). The effects of light-adaptation on rod and cone receptive field organization of monkey ganglion cells. *J. Physiol.*, **192**, 747–760.                                                                 [88]

Graham, C. H., Brown, R. H. & Mote, F. A. (1939). The relation of size of stimulus and intensity in the human eye: I. Intensity thresholds for white light. *J. exp. Psychol.*, **24**, 555–573.                     [164]

Granit, R. (1933). The components of the retinal action potential in mammals and their relation to the discharge in the optic nerve. *J. Physiol.*, **77**, 207–240.                                     [63, 71, 72]

— (1941). A relation between rod and cone substances, based on scotopic and photopic spectra of Cyprinus, Tinca, Anguilla and Testudo. *Acta physiol. Scand.*, **2**, 334–346.                                  [2, 83]

— (1943). The spectral properties of the visual receptors of the cat. *Acta physiol. Scand.*, **5**, 219–229.                                  [83]

— (1947). *Sensory Mechanisms of the Retina*. Oxford University Press, London.                                                                [83]

— (1955). Centrifugal and antidromic effects on ganglion cells of retina. *J. Neurophysiol.*, **18**, 388–411.                            [107]

Granit, R. & Helme, T. (1939). Changes in retinal excitability due to polarization and some observations on the relation between the processes in retina and nerve. *J. Neurophysiol.*, **2**, 556–565.      [63]

Granit, R. & Munsterhjelm, A. (1937). The electrical responses of dark-adapted frogs' eyes to monochromatic stimuli. *J. Physiol.*, **88**, 436–458. [72]

Granit, R. & Svaetichin, G. (1939). Principles and technique of the electrophysiological analysis of colour reception with the aid of microelectrodes. *Upsala Läkaref. Förh.*, **65**, 161–177. [82, 83]

Grassman, H. (1853). Zur Theorie der Farbenmischung. *Ann. Phys., Lpz.*, **89**, 69–84. [144, 214]

s'Gravezande, W. J. (1721). *Physices elementa mathematica, experimentis confirmata.* Lugduni Batavorum: van der Aa (2 vols.). [200, 201]

Green, D. G. (1969). Sinusoidal flicker characteristics of the color-senstitive mechanisms of the eye. *Vision Res.*, **9**, 591–601. [245]

Green, D. G. see also Campbell, F. W.

Green, F. A. see Templeton, W. B.

Grellman, K. -H., Livingston, R. & Pratt, D. (1962). A flash-photolytic investigation of rhodopsin at low temperatures. *Nature.*, **193**, 1258–1260. [12]

Grundfest, H. (1932). The sensitivity of the sun-fish, Lepomis, to monochromatic radiation of low intensities. *J. gen. Physiol.*, **15**, 307–328. [2]

Grüsser-Cornehls, U. see Grüsser, O. -J.

Grüsser, O. -J. (1957). Rezeptorpotentiale einzelner retinaler Zapfen der Katze. *Naturwissenschaften*, **44**, 522–523. [76]

Grüsser, O. -J., Grüsser-Cornehls, U. & Saur, G. (1959). Reaktionen einzelner Neurone im optischen Cortex der Katze nach elektrischer Polarisation des Labyrinths. *Pflügers Arch. ges. Physiol.*, **269**, 593–612. [120]

Grützner, P. (1961). Typische erworbene Farbensinnstörungen bei heredodegenerativen Maculaleiden. *Graefes Arch. Ophthal.*, **163**, 99–116. [230]

Gubisch, R. W. see Campbell, F. W.

Guild, J. (1931). The colorimetric properties of the spectrum. *Phil. Trans.*, **230A**, 149–187. [203]

— (1932). Some problems of visual reception. pp. 1–26 of *Discussion on Vision*. Physical Society, London. [206]

Guillery, R. W. see Colonnier, M.

Guth, S. L., Donley, N. J. & Marrocco, R. T. (1969). On luminance additivity and related topics. *Vision Res.*, **9**, 537–576. [144]

Haft, J. S. & Harman, P. J. (1967). Evidence for central inhibition of retinal function. *Vision Res.*, **7**, 499–501. [109]

Hagins, W. A. (1955). The quantum efficiency of bleaching of rhodopsin in situ. *J. Physiol.*, **129**, 22P–23P. [30]

— (1956). Flash photolysis of rhodopsin in the retina. *Nature*, **177**, 989–990. [29]

— (1958). *Rhodopsin in a mammalian retina.* Ph.D. thesis, Cambridge. [27, 30, 32]

— (1965). Electrical signs of information flow in photoreceptors. *Cold. Spr. Harb. Symp.*, **30**, 403–418. [53]

Hagins, W. A. see also Penn, R. D., Rushton, W. A. H.

Haidinger, W. (1844). Ueber das directe Erkennen des polarisierten Lichts und der Lage der Polarisationsebene. *Ann. Phys., Lpz.*, **63**, 29–39. [141]

Hajos, A. (1969). Verlauf formenspezifischer Farbadaptationen im visuellen System des Menschen. *Psychol. Beiträge*, **11**, 95–114. [147]

von Haller, A. (1763). *Elementa physiologiae corporis humani.* Grasset, Lausanne. [201]

Hallerman, W. see Glees, P.

Hallett, P. E. (1969a). Rod increment thresholds on steady and flashed backgrounds. *J. Physiol.*, **202**, 355–377. [185]

— (1969b). Quantum efficiency and false positive rate. *J. Physiol.*, **202**, 421–436. [197]

Hallett, P. E., Marriott, F. H. C. & Rodger, F. C. (1962). The relationship of visual threshold to retinal position and area. *J. Physiol.*, **160**, 364–373. [163]

Hamasaki, D. I. (1963). The effect of sodium ion concentration on the electroretinogram of the isolated retina of the frog. *J. Physiol.*, **167**, 156–168. [49]

Hamasaki, D. I. (1964). The electroretinogram after application of various substances to the isolated retina. *J. Physiol.*, **173**, 449–458. [68]

Hamasaki, D. I. see also Brindley, G. S., Fujino, T.

Hámori, J. see Szentágothai, J.

Hanitzsch, R. see Byzov, A. L.

Hansen, S. M. see Store J.

Harman, P. J. see Haft, J. S.

Hartline, H. K. (1938). The response of single optic nerve fibers of the vertebrate eye to illumination of the retina. *Amer. J. Physiol.*, **121**, 400–415. [81, 82]

— (1940a). The receptive fields of optic nerve fibers. *Amer. J. Physiol.*, **130**, 690–699. [86]

— (1940b). The effects of spatial summation in the retina on the excitation of the fibers of the optic nerve. *Amer. J. Physiol.*, **130**, 700–711. [86]

Hartridge, H. (1947). The visual perception of fine detail. *Phil. Trans. R. Soc. B.*, **232**, 519–671. [167]

— (1948). Some adaptation effects using yellow stimuli. *J. Physiol.*, **107**, 15P–16P. [214]

Hashimoto, H. see Kaneko, A.

Hashimoto, Y., Murakami, M. & Tomita, T. (1961). Localization of the ERG by aid of histological method. *Jap. J. Physiol.*, **11**, 62–69. [71]

Hassler, R. see Otsuka, R.

Hayashi, R. see Polyak, S. L.

Hayhow, W. R., Webb, C. & Jervie, A. (1960). The accessory optic fiber system in the rat. *J. comp. Neurol.*, **115**, 187–200. [91]

Hayhow, W. R. see also Burke, W.

Hécaen, H., Penfield, W., Bertrand, C. & Malmo, R. (1956). The syndrome of apractognosia due to lesions of the minor cerebral hemisphere. *Arch. Neurol. Psychiat.*, **75**, 400–434. [131]

Hecht. S. & Pickels, E. G. (1938). The sedimentation constant of visual purple. *Proc. Nat. Acad. Sci.*, **24**, 172–176. [4]

Hecht, S. & Shlaer, S. (1936). The relation between intensity and critical frequency for different parts of the spectrum. *J. gen. Physiol.*, **19**, 965–977. [173]

Hecht, S., Shlaer, S. & Pirenne, M. (1942). Energy, quanta and vision. *J. gen. Physiol.*, **25**, 819–840. [186, 189]

Hecht, S. see also Lamar, E. S.

Heller, J. (1968). Structure of visual pigments. I. Purification, molecular weight and composition of bovine visual pigment. *Biochemistry*, **7**, 2906–2913. [4, 5, 21]

— (1969). Comparative study of a membrane protein. Characterization of bovine, rat and frog visual pigment. *Biochem.*, **8**, 675–678. [5, 21]

Helme, T. see Granit, R.

von Helmholtz, H. (1852). Über die Theorie der zusammengesetzten Farben. *Ann. Phys. Lpz.*, **87**, 45–66. [203]

— (1863). *Die Lehre von den Tonempfindungen als physiologische Grundlage für die Theorie der Musik.* Vieweg, Braunschweig. [201]

— (1867). *Handbuch der physiologischen Optik.* 1st edn. Voss, Leipzig. [145, 216]

— (1896). *Handbuch der phyiologischen Optik*, 2nd edn. Voss, Leipzig. [53]

Helms, A. & Prehn, R. (1958). Empfindlichkeitsänderungen des dunkeladaptierten menschlichen Auges bei monocularer Adaptation in Abhängigkeit vom Helladaptationsniveau. *v. Graefes Arch. Ophthal.*, **106**, 285–289. [109]

Helson, H. (1943). Some factors and implications of colour constancy. *J. opt. Soc. Amer.*, **33**, 555–567. [152]

Hendley, C. D. see Lamar, E. S.

Henry, G. H., Cole, B. L. & Nathan, J. (1964). The inheritance of congenital tritanopia, with the report of an extensive pedigree. *Ann. Hum. Gen.*, **27**, 219–231. [228]

Henry, G. H. see also Rushton, W. A. H.

Henschen, S. E. (1898). Über Localisation innerhalb des äusseren Knieganglions. *Neurol. Centralblatt*, **17**, 194–199. [99]

Hering, E. (1875). Zur Lehre vom Lichtsinne. VI. Grundzüge einer theorie des Farbensinnes. *SB K. Akad. Wiss Wien. Math. naturwiss. K.*, **70**, 169–204. [206]

Heron, W. see Burns, B. D.

Hill, R. M. see Barlow, H. B., Horn, G.

Hillman, B. M. (1958). Relationship between stimulus size and threshold intensity in the fovea measured at four exposure times. *J. opt. Soc. Amer.*, **48**, 422–428. [164]

Hirsch, J. see Calvet, J.

Hobson, R. R. see Weinstein, G. W.

Holden, A. L. (1968). Antidromic activation of the isthmo-optic nucleus. *J. Physiol.*, **197**, 183–198.                                    [106]

Holmes, G. (1918). Disturbances of vision by cerebral lesions. *Brit. J. Ophthal.*, **2**, 353–384.                                    [110, 112]

Hooke, Robert (1665). *Micrographia: or some physiological descriptions of minute bodies made by magnifying glasses, with observations and inquiries thereupon.* Martyn & Allestry, London.                                    [201]

Horn, G. (1965). The effect of somaesthetic and photic stimuli on the activity of units in the striate cortex of unanaesthetized unrestrained cats. *J. Physiol.*, **179**, 263–277.                                    [120]

Horn, G. & Hill, R. M. (1966). Responsiveness to sensory stimulation of units in the superior colliculus and subjacent tectotegmental regions of the rabbit. *Exp. Neurol.*, **14**, 199–223.                                    [94]

— (1969). Modifications of receptive fields of cells in the visual cortex occurring spontaneously and associated with bodily tilt. *Nature*, **221**, 186–188.                                    [120, 126]

Hotta, T. & Kameda, K. (1964). Effect of midbrain stimulation on a visual response. *Exp. Neurol.*, **9**, 127–136.                                    [94, 104]

Hubbard, R. (1954). The molecular weight of rhodopsin and the nature of the rhodopsin-digitonin complex. *J. gen. Physiol.*, **37**, 381–399.                                    [4, 13, 23]

— (1956). Retinene isomerase. *J. gen. Physiol.*, **39**, 935–962.    [18, 33]

— (1958). The thermal stability of rhodopsin and opsin. *J. gen. Physiol.*, **42**, 259–280.                                    [4]

— (1969). Absorption spectrum of rhodopsin: 280 nm absorption band. *Nature*, **221**, 435–437.                                    [4, 13]

Hubbard, R., Bownds, D. & Yoshizawa, T. (1965). The chemistry of visual photoreception. *Cold Spr. Harb. Symp. Quant. Biol.*, **30**, 301–315.                                    [5, 30]

Hubbard, R. & Colman, A. D. (1959). Vitamin-A content of the frog eye during light and dark adaptation. *Science*, **130**, 977–978.    [35]

Hubbard, R. & Dowling, J. E. (1962). Formation and utilization of 11-*cis* vitamin A by the eye tissues during light and dark adaptation. *Nature*, **193**, 341–343.                                    [35]

Hubbard, R. & Kropf, A. (1958). The action of light on rhodopsin. *Proc. Nat. Acad. Sci.*, **44**, 130–139.                                    [22, 30]

— (1959). Chicken lumi- and meta-iodopsin. *Nature*, **183**, 448–450. [40]

Hubbard, R. & St George, R. C. C. (1958). The rhodopsin system of the squid. *J. gen. Physiol.*, **41**, 501–528.                                    [30]

Hubbard, R. & Wald, G. (1951). The mechanism of rhodopsin synthesis. *Proc. Nat. Acad. Sci.*, **37**, 69–79.                                    [15]

— (1952). Cis-trans isomers of vitamin A and retinene in the rhodopsin system. *J. gen. Physiol.*, **36**, 269–315.                                    [15]

Hubbard, R. see also Matthews, R. G.

Hubel, D. H. (1958). Cortical unit responses to visual stimuli in non-anesthetized cats. *Amer. J. Ophthal.*, **46**, 110–121.                                    [117]

— (1960). Single unit activity in lateral geniculate body and optic tract of unrestrained cats. *J. Physiol.*, **150**, 91–104.                                    [104]

Hubel, D. H. & Wiesel, T. N. (1959). Receptive fields of single neurones in the cat's striate cortex. *J. Physiol.*, **148**, 574–591. [117]

— (1960). Receptive fields of optic nerve fibres in the spider monkey. *J. Physiol.*, **154**, 572–580. [84, 88]

— (1961). Integrative action in the cat's lateral geniculate body. *J. Physiol.*, **155**, 385–398. [103]

— (1962). Receptive fields, binocular interaction and functional architecture in the cat's striate cortex. *J. Physiol.*, **160**, 106–154. [117, 119]

— (1963). Shape and arrangement of columns in cat's striate cortex. *J. Physiol.*, **165**, 559–568. [118]

— (1965). Receptive fields and functional architecture in two non-striate visual areas (18 and 19) of the cat. *J. Neurophysiol.*, **28**, 229–289. [111, 118]

— (1967). Cortical and callosal connections concerned with the vertical meridian of visual fields in the cat. *J. Neurophysiol.*, **30**, 1561–1573. [127]

— (1968). Receptive fields and functional architecture of monkey striate cortex. *J. Physiol.*, **195**, 215–243. [117, 118, 119]

— (1969a). Anatomical demonstration of columns in the monkey striate cortex. *Nature*, **221**, 747–750. [118]

— (1969b). Visual area of the lateral suprasylvian gyrus (Clare-Bishop area) of the cat. *J. Physiol.*, **202**, 251–260. [111]

Hubel, D. H. see also Wiesel, T. N.

Hughes, G. W. & Maffei, L. (1965). On the origin of the dark discharge of retinal ganglion cells. *Arch. ital. Biol.*, **103**, 45–59. [81]

Humphrey, N. K. & Weizkrantz, L. (1967). Vision in monkeys after removal of the striate cortex. *Nature*, **215**, 595–597. [92]

Hunter, J. & Ingvar, D. H. (1955). Pathways mediating metrazol-induced irradiation of visual impulses. *Electro-enceph. clin. Neurophysiol.*, **7**, 39–60. [103]

Hurvich, L. M. & Jameson, D. (1957). An opponent-process theory of color vision. *Psychol.*, *Rev.*, **64**, 384–404. [209]

Ikeda, H. & Ripps, H. (1966). The electroretinogram of a cone-monochromat. *Arch. Ophthal.*, **75**, 513–517. [229]

Ikeda, H. see also Arden, G. B.

Ikeda, M. (1965). Temporal summation of positive and negative flashes in the visual system. *J. opt. Soc. Amer.*, **55**, 1527–1534. [177]

Ikeda, M. & Fujii, T. (1966). Diphasic nature of the visual response as inferred from the summation index of n flashes. *J. opt. Soc. Amer.*, **56**, 1129–1132. [177]

Ikeda, M. see also Boynton, R. M.

Ingvar, D. H. see Hunter, J.

Irreverre, F. see Shichi, H.

Ives, H. E. (1922). Critical frequency relations in scotopic vision. *J. opt. Soc. Amer.*, **6**, 254–268. [174]

Iawi, E. see Ettlinger, G.

Iwama, K., Sakakura, H. & Kasamatsu, T. (1965). Presynaptic inhibi-

L

tion in the lateral geniculate body induced by stimulation of the cerebral cortex. *Jap. J. Physiol.*, **15**, 311–322.     [102]

Iwama, K. see also Sakakura, H.

Jacobson, J. H. & Gestring, G. F. (1958). Centrifugal influence upon the electroretinogram. *Arch. Ophthal. (Chicago)*, **60**, 295–302.     [108]

Jaeger, W. (1955). Tritoformen angeborener und erworbener Farbensinnstörungen. *Farbe*, **4**, 197–215.     [229]

James, M. see Warrington, E. K.

Jameson, D. see Hurvich, L. M.

von Jancsó, N. & von Jancsó, H. (1936). Fluoreszenzmikroskopische Beobachtung der reversiblen Vitamin-A-Bildung in der Netzhaut während des Sehaktes. *Biochem. Z.*, **287**, 289–290.     [35]

Jasper, H. see Penfield, W.

Jean, J. N. & O'Brien, B. (1949). Microwave test of a theory of the Stiles and Crawford effect. *J. opt. Soc. Amer.*, **39**, 1057.     [253]

Jeanne, M. see Fiorentini, A.

Jeremy, D. see Bishop, P. O.

Jervie, A. (Sefton, Mrs. A. J.) see Burke, W., Hayhow, W. R. and her married name.

Johnson, E. P. see also Armington, J. C.

Joliffe, C. L. see Sperling, H. G.

Jolly, W. A. see Einthoven, W.

Judd, D. B. (1949). Response functions for types of vision according to the Müller theory. *J. Research Natl. Bur. Stds.*, **42**, 1–15.     [209]

Jung, R., von Baumgarten, R. & Baumgartner, G. (1952). Mikroableitungen von einzelnen Nervenzellen im optischen Cortex der Katze: Die Lichtaktivierten B-Neurone. *Arch. f. Psychiatr. u. Nervenkr.*, **189**, 521–539.     [116]

Jung, R. Kornhuber, H. H. & da Fonseca, J. S. (1963). Multisensory convergence on cortical neurons: neuronal effects of visual, acoustic and vestibular stimuli in the superior convolutions of the cat's cortex. *Prog. Brain Res.*, **1**, 207–240.     [120]

Kaiser, P. K. see Boynton, R. M.

Kalsbeck, J. E. see Ettlinger, G.

Kameda, K. see Hotta, T.

Kandel, G. see Boynton, R. M.

Kaneko, A. & Hashimoto, H. (1967). Recording site of the single cone response determined by an electrode marking technique. *Vision Res.*, **7**, 847–851.     [48, 69, 75]

— (1968). Localization of spike-producing cells in the frog retina. *Vision Res.*, **8**, 259–262.     [75, 80]

— (1969). Electrophysiological study of single neurons in the inner nuclear layer of the carp retina. *Vision Res.*, **9**, 37–55.     [80]

Kaneko, A. see also Murakami, M., Tomita, T.

Kasamatsu, T. see Iwama, K.

Kato, E. see Suzuki, H.

Kawamura, H. & Marchiafava, P. L. (1968). Excitability changes along visual pathways during tracking eye movements. *Arch. ital. Biol.*, **106**, 141–156. [103]

Kimbel, R. L. see Poincelot, R. P.

Kinney, Jo Ann S. (1967). Color induction using asynchronous flashes. *Vision Res.*, **7**, 229–318. [152]

Kinsbourne, M. see Warrington, E. K.

Klüver, H. & Bucy, P. C. (1939). Preliminary analysis of functions of the temporal lobes in monkeys. *Arch. Neurol. Psychiatr.*, **42**, 979–1000. [129]

Kohlrausch, A. (1918). Die Netzhautströme der Wirbeltiere in Abhängigkeit von der Wellenlänge des Lichtes und dem Adaptationszustand des Auges. *Arch. Anat. Physiol. (Lpz.).*, **1918**, 195–241. [73]

— (1922). Untersuchungen mit farbigen Schwellenprüflichtern über den Dunkeladaptationsverlauf des normalen Auges. *Pflügers Arch. ges. Physiol.*, **196**, 113–117. [177]

— (1931). Tagessehen, Dämmersehen, Adaptation. Pp. 1499–1594 of *Handbuch der normalen und pathologischen Physiologie.* Bd. 12/2. Springer, Berlin. [136, 177, 178]

König, A. (1894). Über den menschlichen Sehpurpur und seine Bedeutung für das Sehen. *S. B. Akad. Wiss. Berlin*, **1894**, 577–598. [240, 257]

— (1896). Quantitative Bestimmungen an complementären Spectralfarben. *S. B. Akad. Wiss. Berlin*, **1896**, 945–949. [204, 216, 258]

König, A. & Dieterici, C. (1886). Die Grundempfindungen und ihre Intensitäts-Vertheilung im Spectrum. *S. B. Akad. Wiss. Berlin*, **1886**, 805–829. [203, 223]

König, A. & Zumft, J. (1894). Über die lichtempfindliche Schicht in der Netzhaut des menschlichen Auges. *S. B. Akad. Wiss. Berlin*, **1894**, 439–442. [139]

Kornhuber, H. H. see Jung, R.

Köttgen, E. & Abelsdorff, G. (1896). Absorption und Zersetzung des Sehpurpurs bei den Wirbeltieren. *Z. Psychol. Physiol. Sinnesorg.*, **12**, 161–184. [36]

Kozak, W., Rodieck, R. W. & Bishop, P. O. (1965). Responses of single units in lateral geniculate nucleus of cat to moving visual patterns. *J. Neurophysiol.*, **28**, 19–47. [103]

Kozak, W. see also Bishop, P. O.

Krause, F. & Schum, H. (1931). *Neue Deutsche Chirurgie.* Band **49a**, pp 482–486. Stuttgart: Enke. [124]

von Kries, J. (1878). Beitrag zur Physiologie der Gesichtsempfindungen *Arch. Anat. Physiol. Lpz.*, **1878**, 504–524. [216]

— (1896). Über die Funktion der Netzhautstäbchen. *Z. Psychol. Physiol., Sinnesorg.*, **9**, 81–123. [204, 215, 216, 258]

— (1905). *Handbuch der Physiologie des Menschen*, ed. W. Nagel, Bd. **3**, S. 211. Springer, Berlin. [246]

Krieger, H. P. See Pasik, P.

Krinsky, N. I. (1958a). The lipoprotein nature of rhodopsin. *Arch. Ophthal., Chicago*, **60**, 688–694. [4]

Krinsky, N. I. (1958b). The enzymatic esterification of vitamin A. *J. biol. Chem.*, **232**, 881–894. [35]

Kropf, A. (1967). Intramolecular energy transfer in rhodopsin. *Vision Res.*, **7**, 811–818. [12, 13]

Kropf, A. see also Hubbard, R.

Kuffler, S. W. (1953). Discharge patterns and functional organization of mammalian retina. *J. Neurophysiol.*, **16**, 37–68. [87, 177]

Kuffler, S. W., Fitzhugh, R. & Barlow, H. B. (1957). Maintained activity in the cat's retina in light and darkness. *J. gen. Physiol.*, **40**, 683–702. [81]

Kuffler, S. W. see also Barlow, H. B.

Kühne, W. (1878a). Zur Photochemie der Netzhaut. *Untersuchungen. physiol. Inst. Univ. Heidelberg*, **1**, 1–14. [1, 33]

— (1878b). Über den Sehpurpur. *Untersuchungen physiol. Inst. Univ. Heidelberg*, **1**, 15–103. [1, 26]

— (1882a). Beobachtungen an der frischen Netzhaut des Menschen. *Untersuchungen physiol. Inst. Univ. Heidelberg*, **2**, 69–80. [1]

— (1882b). Fortgesetzte Untersuchungen über die Retina und die Pigmente des Auges. *Untersuchungen physiol. Inst. Univ. Heidelberg*, 2, 89–132. [1]

Kühne, W. & Steiner, J. (1880). Über das electromotorische Verhalten der Netzhaut. *Untersuchungen physiol. Inst. Univ. Heidelberg*, **3**, 327–377. [61]

Kühne, W. see also Ayres, W. C., Ewald, A.

Kuhns, M. P. see Brown, J. L.

Kulikowski, J. J. see Campbell, F. W.

Kuman, E. A. & Strebitskiï, V. G. (1968). The interaction of visual and auditory signals in the lateral geniculate body of the rabbit's brain (in Russian). *Zhurn. vyssh. nerv. deiat.*, **18**, 507–513. [104]

Lamar, E. S., Hecht, S., Shlaer, S. & Hendley, C. D. (1947). Size, shape and contrast in detection of targets by daylight vision: I. Data and analytical description. *J. opt. Soc. Amer.*, **37**, 531–545. [164, 165]

de Lange, H. (1958). Research into the dynamic nature of the human fovea-cortex systems with intermittent and modulated light. I. Attenuation characteristics with white and colored light. *J. opt. Soc. Amer.*, **48**, 777–784. [174]

Lashley, K. S. (1943). Loss of the maze habit after occipital lesions in blind rats. *J. comp. Neurol.*, **79**, 431–462. [126]

— (1948). Effects of destroying the visual 'associative areas' of the monkey. *Genet. Psychol. Monogr.*, **37**, 107–166. [130]

Latour, P. L. (1962). Visual threshold during eye movements. *Vision Res.*, **2**, 261–262. [186]

Lee, G. B. see Alpern, M.

Le Grand, Y. (1935). Sur la mesure de l'acuité visuelle au moyen de franges d'interférence. *C. R. Acad. Sci., Paris*, **200**, 490–491. [166, 168]

Leicester, J. & Stone, J. (1967). Ganglion, amacrine and horizontal cells of the cat's retina. *Vision Res.*, **7**, 695–705. [57]

Lennox, W. G. see Wolff, H. G.

Leonardo da Vinci (1651). *Trattato della Pittura*. Langlois, Paris. [200]

Leontyev, V. G. see Etingov, R. N.

Lettvin. J. Y. (1965). Contribution to discussion on the early receptor potential. *Cold. Spr. Harb. Symp.*, **30**, 501–502.                [32]

Lettvin, J. Y. see also Maturana, H. R.

Levick, W. R. (1967). Receptive fields and trigger features of ganglion cells in the visual streak of the rabbit retina. *J. Physiol.*, **188**, 285–307. [88]

Levick, W. R., Oyster, C. W. & Takahashi, E. (1969). Rabbit lateral geniculate nucleus: sharpener of directional information. *Science*, **165**, 712–714.                [104]

Levick, W. R. see also Barlow, H. B., Bishop, P. O., Ogawa, T.

Levinson, J. (1960a). Fusion of complex flicker. *Science*, **130**, 919–921. [174]

— (1960b). Fusion of complex flicker II. *Science*, **131**, 1438–1440. [174]

Levinson, J. see also Campbell, F. W.

Lewin, W. see Brindley, G. S.

Lewis, D. M. (1957). Regeneration of rhodopsin in the albino rat. *J. Physiol.*, **136**, 624–631.                [35]

Lewis, M. S. see Shichi, H.

Liaudansky, L. H. see Mitchell, R. T.

Liberman, E. A. (1957). The character of the information entering the brain from two retinal receptors of the frog along a single nerve fibre (in Russian). *Biofizika*, **2**, 427–430.                [84]

Liebman, P. A. & Entine, G. (1964). Sensitive low-light level microspectrophotometer: detection of photo-sensitive pigments of retinal cones. *J. opt. Soc. Amer.*, **54**, 1451–1459.                [39, 44]

Liss, L. see Wolter, J. R.

Livingston, R. see Grellman, K. -H.

Lomonosov, M. V. (1757). *Lecture on the origin of light, presenting a new theory of colours* (in Russian). St Petersburg: Academy of Sciences. The passage quoted is translated from p. 335 of the 1952 Moscow edition of Lomonosov's collected works.                [201]

Love, R. M. see Collins, F. D.

Lozano, R. D. & Palmer, D. A. (1967). The additivity of large-field colour matching functions. *Vision Res.*, **7**, 929–937.                [216]

Ludvigh, E. & McCarthy, E. F. (1938). Absorption of visible light by the refractive media of the human eye. *Arch. Ophthal. (Chicago)*, **20**, 37–51.                [187]

Luizov, A. V. see Bulanova, K. N.

Lythgoe, R. J. (1932). The measurement of visual acuity. M.R.C. Spec. Rep., No. **173**.                [168]

— (1937). The absorption spectra of visual purple and of indicator yellow. *J. Physiol.*, **89**, 331–358.                [6, 20]

— (1940). The mechanism of dark adaptation: a critical resumé. *Brit. J. Ophthal.*, **24**, 21–43.                [180]

Lythgoe, R. J. & Quilliam, J. P. (1938a). The thermal decomposition of visual purple. *J. Physiol.*, **93**, 24–38.                                    [4]
— (1938b). The relation of transient orange to visual purple and indicator yellow. *J. Physiol.*, **94**, 399–410.                              [6, 8]
Lythgoe, R. J. see also Broda, E. E., Dartnall, H. J. A., Goodeve, C. F., Schneider, E.
Macadam, D. L. (1942). Visual sensitivities to color differences in daylight. *J. opt. Soc. Amer.*, **32**, 247–274.                            [231]
— (1949a). Measurement of the influence of local adaptation on color matching.                                                   [151, 245, 247]
— (1949b). Color discrimination and the influence of color contrast on acuity. *Docum. ophthal.*, **3**, 214–233.                            [231]
— (1956). Chromatic adaptation. *J. opt. Soc. Amer.*, **46**, 500–513.
                                                             [151, 245, 247]
— see also Brown, W. R. J.
McCarthy, E. F. see Ludvigh, E.
McCollough, C. (1965). Color adaptation of edge-detectors in the human visual system. *Science*, **149**, 1115–1116.                      [147]
McCree, K. J. (1960). Small-field tritanopia and the effect of voluntary fixation. *Optica Acta.*, **7**, 317–323.                             [244]
McCulloch, W. S. see Matwana, H. R.
McGill, J. J. (1964). Organization within the central and centrifugal fibre pathways of the avian visual system. *Nature*, **204**, 395–396. [106]
Mach, E. (1865). Über die Wirkung der räumlichen Vertheilung des Lichtreizes auf die Netzhaut. *S. B. Kais. Akad. Wiss. Wien (Math.-Nat. Kl.).*, **52**, 303–322.                                        [153]
McIlwain, J. T. (1964). Receptive fields of optic tract axons and lateral geniculate cells: peripheral extent and barbiturate sensitivity. *J. Neurophysiol.*, **27**, 1154–1173.                             [87]
— (1966). Some evidence concerning the physiological basis of the periphery effect in the cat's retina. *Exp. Brain Res.*, **1**, 265–271. [87]
Mackay, D. M. (1968). Evoked potentials reflecting interocular and monocular suppression. *Nature*, **217**, 81–83.                       [116]
McKeown, T. see Clark, W. E. Le Gros.
McLeod, J. G. see Bishop, P. O.
MacNichol, E. J. & Svaetichin, G. (1958). Electric responses from the isolated retinas of fishes. *Amer. J. Ophthal.*, **46**, 26–40.     [47, 76]
MacNichol, E. F. see also Marks, W. B., Wagner, H. G.
Maffei, L. & Poppele, R. E. (1968). Transient and steady state electroretinal responses. *Vision Res.*, **8**, 229–246.                      [71]
Maffei, L. see Hughes, G. W.
Magni, F. see Angel, A.
Magoun, H. W. see Ranson, S. W.
Major, D. see Brown, J. E., Dowling, J. E.
Malmo, R. see Hécaen, H.
Marchiafava, P. L. see Kawamura, H.
Mariotte, E. (1681). *Traité des couleurs*. Paris.              [200, 201, 202]
Marks, W. B. (1963). *Difference spectra of the visual pigments in single*

*goldfish cones*. Ph.D. thesis, Johns Hopkins University, Baltimore.
[39, 44]

Marks, W. B. (1965). Visual pigments of single goldfish cones. *J. Physiol.*,
**178**, 14–32. [39, 44]

Marks, W. B., Dobelle, W. H. & MacNichol, E. F. (1964). Visual pigments of single primate cones. *Science*, **143**, 1181–1183. [39, 44, 45,
221]

Marriott, F. H. C. see Pirenne, M. H., Hallett, P. E.

Marrocco, R. T. see Guth, S. L.

Marshall, W. H. (1949). Excitability cycle and interaction in geniculate-
striate system of cat. *J. Neurophysiol.*, **12**, 277–288. [102]

Marshall, W. H., Talbot, S. A. & Ades, H. W. (1943). Cortical response of
the anesthetized cat to gross photic and electrical afferent stimulation.
*J. Neurophysiol.*, **6**, 1–15. [112, 115]

Marshall, W. H. see also Talbot, S. A.

Marzi, C. A. see Mascetti, G. G.

Mascetti, G. G., Marzi, C. A. & Berlucchi, G. (1969*a*). Sympathetic
influences on the dark-discharge of the retina in the freely moving
cat. *Arch. ital., Biol.*, **107**, 158–166. [108]

— (1969*b*). Changes in resting activity of retinal ganglion cells produced
by electrical stimulation of the cervical sympathetic trunk. *Arch.
ital. biol.*, **107**, 167–174. [108]

Massion, J. & Meulders, M. (1961). Les potentiels evoqués visuels et
auditifs du centre médian et leurs modifications apres décortication.
*Arch. internat. Physiol Bioch.*, **69**, 26–29. [95]

Matthews, B. H. C. see Adrian, E. D.

Matthews, R. see Adrian, E. D.

Matthews, R. G., Hubbard, R., Brown, P. K. & Wald, G. (1963).
Tautomeric forms of metarhodopsin. *J. gen. Physiol.*, **47**, 215–240.
[10, 11, 21, 29]

Maturana, H. R. Lettvin, J. Y., McCulloch, W. S. & Pitts, W. H. (1960).
Anatomy and physiology of vision in the frog (*Rana pipiens*). *J. gen.
Physiol.*, **43**, 129–176. [88]

Maxwell, J. C. (1856). On the unequal sensibility of the Foramen
Centrale to light of different colours. *Rep. Brit. Assoc.*, **1856**, 12.
[140]

Mayer, T. (1758). Von Messung der Farben durch Hülfe der Vermisch-
ung. *Göttingische gelehrte Anzeigen*, **1758**, 1385–1389. [201]

Meikle, T. H. see Sprague, J. M.

Mertens, J. J. (1956). Influence of knowledge of target location upon the
probability of observation of peripherally observable test flashes.
*J. opt. Soc. Amer.*, **46**, 1069–1070. [109]

Meulders, M. & Godfraind, J. M. (1969). Influence du réveil d'origine
réticulaire sur l'étendue des champs visuels des neurones de la
région genouillée chez le chat avec cerveau intact ou avec cerveau
isolé. *Exp. Brain Res.*, **9**, 201–220. [104]

Meulders, M. see also Massion, J., Godfraind, J. M.

Meyer, J.S., Fang, H. C. & Denny-Brown, D. (1954). Polarographic study of cerebral collateral circulation. *Arch. Neurol. Psychiat.*, **72**, 296–312.     [123]

Michael, C. R. (1968). Receptive fields of single optic nerve fibres in a mammal with an all-cone retina. III: Opponent color units. *J. Neurophysiol.*, **31**, 268–282.     [85]

Michaels, J. see Missotten, L.

Millar, P. G. see Poincelot, R. P.

Miller, G. L. see Arden, G. B.

Minkowski, M. (1920). Über den Verlauf, die Endigung und die zentrale Repräsentation von gekreuzten und ungekreuzten Sehnervenfasern bei einigen Säugetieren und beim Menschen. *Schweiz. Arch. Neurol. Psychiat.*, **6**, 201–252.     [98]

Mishkin, M. (1954). Visual discrimination performance following partial ablations of the temporal lobe: II. Ventral surface vs. hippocampus. *J. comp. physiol. Psychol.*, **47**, 187–193.     [129]

— (1966). Chap. 4 of *Frontiers in physiological psychology*. New York: Academic Press.     [130]

Mishkin, M. & Pribram, K. H. (1954). Visual discrimination performance following partial ablations of the temporal lobe: I. Ventral vs. lateral. *J. comp. physiol. Psychol.*, **47**, 14–20.     [129]

Mishkin, M. see also Ettlinger, G., Weiskrantz, L.

Missotten, L., Appelmans, M. & Michiels, J. (1963). L'ultrastructure des synapses des cellules visuelles de la rétine humaine. *Bull. Mém. Soc. Franç. Ophtal.*, **76**, 59–82.     [57, 59]

Mitchell, R. T. & Liaudansky, L. H. (1955). Effect of differential adaptation of the eyes upon threshold sensitivity. *J. opt. Soc. Amer.*, **45**, 831–834.     [109]

Mommaerts, W. F. H. M. see Crescitelli, F.

von Monakow, C. (1889). Experimentelle und pathologisch-anatomische Untersuchungen über die optischen Centren und Bahnen. *Arch. Psychiatr. Nervenkr.*, **20**, 714–787.     [106]

Moreland, J. D. (1968). On demonstrating the blue arcs phenomenon. *Vision Res.*, **8**, 99–107.     [159]

Moreland, J. D. & Cruz, A. (1959). Colour perception with the peripheral retina. *Optica Acta.*, **6**, 117–151.     [204]

Morton, H. B. see Cobb, W. A.

Morton, R. A. & Goodwin, T. W. (1944). Preparation of retinene *in vitro. Nature, Lond.*, **153**, 405–406.     [3, 19]

Morton, R. A. & Pitt, G. A. J. (1955). Studies on rhodopsin. 9. pH and the hydrolysis of indicator yellow. *Biochem., J.* **59**, 128–134.     [19, 20]

Morton, R. A. see also Ball, S., Collins, F. D.

Mote, F. A. & Forbes, L. M. (1957). Changing pre-exposure and dark adaptation. *J. opt. Soc. Amer.*, **47**, 287–290.     [184]

Mote, F. A. see also Graham, C. H.

Motokawa, K. & Ebe, M. (1954). Antidromic stimulation of optic nerve and photosensitivity of cat retina, *J. Neurophysiol.*, **17**, 364–373.     [107]

Motokawa, K. see also Oikawa, T.

Müller, G. E. (1896). Zur Psychophysik der Gesichtsempfindungen. Z. Psy. Physiol., Sinnesorg., 10, 1–82 and 321–413.          [209]

Müller, H. (1851). Zur Histologie der Netzhaut. Z. wiss. Zool., 3, 234–237.          [1]

— (1854). S. B. Würzb. phys.-med. Gesellsch., May 27th and Nov. 4th, 1854. Cited by König and Zumft (1894). q.v.          [139]

Munsterhjelm, A. see Granit, R.

Munz, F. W. (1958). Photosensitive pigments from the retinae of certain deep-sea fishes. J. Physiol., 140, 220–235.          [36]

Murakami, M. & Kaneko, A. (1966). Subcomponents of PIII in cold-blooded vertebrate retinae. Nature, 210, 103–104.          [50, 69, 70]

Murakami, M. see also Brown, K. T., Hashimoto, Y., Tomita, T.

Myers, R. E. (1955). Interocular transfer of pattern discrimination in cats following section of crossed optic fibers. J. comp. physiol. Psychol., 48, 470–473.          [128]

— (1959). Localization of function in the corpus callosum. Arch. Neurol., 1, 74–77.          [128]

— (1962). Commissural connexions between occipital lobes of the monkey. J. comp. Neurol., 118, 1–16.          [127]

— (1964). Visual deficits after lesions of brain stem tegmentum in cats. Arch. Neurol., 11, 73–90.          [93]

— see also Black, P., Sperry, R. W.

Naeve, H. see Glees, P.

Nagel, W. see Abelsdorff, G.

Naka, K. I. & Rushton, W. A. H. (1966a). An attempt to analyse colour reception by electrophysiology. J. Physiol., 185, 555–586.          [2]

— (1966b). S-potentials from luminosity units in the retina of fish (Cyprinidae). J. Physiol., 185, 587–599.          [76]

— (1967). The generation and spread of s-potentials in fish (Cyprinidae). J. Physiol., 192, 437–461.          [78]

— (1968). S-potential and dark adaptation in fish. J. Physiol., 194, 259–269.          [183]

Nathan, J. see Henry, G. H.

Nechaeva, N. V. see Brodskii, V. Ya., Utina, I. A.

Nelson, J. H. see Wright, W. D.

Newcombe, F. see Whitty, C. W. M.

Newhall, S. M. (1937). The constancy of the blue arc phenomenon. J. opt. Soc. Amer., 27, 165–176.          [159]

Newton, Sir Isaac (1704). Opticks. Sam. Smith & Benj. Walford, London.          [200]

Noell, W. K. (1954). The origin of the electroretinogram. Amer. J. Ophthal., 38, 78–90.          [72]

Nosaki, H. see Toyoda, J.

Ogawa, T., Bishop, P. O. & Levick, W. R. (1966). Temporal characteristics of responses to photic stimulation by single ganglion cells in the unopened eye of the cat. J. Neurophysiol., 29, 1–30.          [88]

Ogawa, T. see also Oikawa, T.

Ogden, T. E. & Brown, K. T. (1964). Intraretinal responses of the cynamolgus (sic) monkey to electrical stimulation of the optic nerve and retina. *J. Neurophysiol.*, **27**, 682–705.                    [107]

Oikawa, T., Ogawa, T. & Motokawa, K. (1959). Origin of the so-called cone action potential. *J. Neurophysiol.*, **22**, 102–111.            [47]

O'Leary, J. S. (1940). A structural analysis of the lateral geniculate nucleus of the cat. *J. comp. Neurol.*, **73**, 405–430.          [98, 100]

Østerberg, G. (1935). Topography of the layer of rods and cones in the human retina. *Acta ophthal. Kbh., Suppl.* 6.                [258]

Ostrovskiĭ, M. A. & Dettmar, P. (1968). The influence of ouabain on the electroretinogram of the isolated superfused retina of the frog (in Russian). *Biofizika*, **11**, 724–726.                      [49]

Ostrovskiĭ, M. A. & Trifonov, Yu. A. (1967). The reaction of horizontal cells of the retina of the carp when Na-K activated ATPase is inhibited with ouabain (in Russian). *Biofizika*, **12**, 1037–1042.   [79]

Ostroy, S. E., Erhardt, F. & Abrahamson, E. W. (1966). The sequence of intermediates in the thermal decay of cattle metarhodopsin in vitro. *Biochim. Biophys. Acta.*, **112**, 265–277.               [21]

Otsuka, R. & Hassler, R. (1962). Über Aufbau und Gliederung der corticalen Sehsphäre bei der Katze. *Arch. Psychiatr. Nervenkr.*, **203**, 212–234.                                        [111]

Oyster, C. W. see Levick, W. R.

Padgham, C. A. (1968). Measurements of the colour sequences in positive visual after-images. *Vision Res.*, **8**, 939–949.            [149]

Pak, W. L. (1965). Some properties of the early electrical response in the vertebrate retina. *Cold. Spr. Harb. Symp.*, **30**, 493–499.   [31]

Pak, W. L., Rozzi, V. P. & Ebrey, T. G. (1967). Effect of changes in the chemical environment of the retina on the two components of the early receptor potential. *Nature*, **214**, 109–110.            [32]

Palay, S. L. see Peter, A.

Palmer, D. A. see Lozano, R. D.

Palmer, G. (1777). *Theory of colours and vision.* S. Leacroft, London.
                                                              [202]

Pantle, A. J. & Sekuler, R. W. (1968). Velocity-sensitive elements in human vision: initial psychophysical evidence. *Vision Res.*, **8**, 445–450.                                          [146]

Parry, H. B., Tansley, K. & Thomson, L. C. (1953). The electro-retinogram of the dog. *J. Physiol.*, **120**, 28–40.          [62, 72]

Pasik, P. & Pasik, T. (1968). Further studies on extrageniculostriate vision in the monkey. *Trans. Amer. neurol. Assoc.*, **93**, 262–264.    [92]

Pasik, P., Pasik, T. & Krieger, H. P. (1959). Effects of cerebral lesions on optokinetic nystagmus in monkeys. *J. Neurophysiol.*, **22**, 297–304.
                                                              [93]

Pasik, P., Pasik, T. & Schilder, P. (1969). Extrageniculostriate vision in the monkey: discrimination of luminous flux-equated figures. *Exp. Neurol.*, **24**, 421–437.                          [92]

Pasik, T., Pasik, P. & Bender, M. B. (1966). The superior colliculi and eye movements. *Arch. Neurol.*, **15**, 420–436.            [93]

Pasik, T. see also Pasik, P.

Pautler, E. L. see Tomita, T.

Pedler, C. M. H. & Tilly, R. (1967). The fine structure of photoreceptive discs. *Vision Res.*, **7**, 829–836.                    [24]

Penfield, W. & Jasper, H. (1954). *Epilepsy and the functional anatomy of the human brain.* Churchill, London.        [121, 122, 123, 124, 130]

Penfield, W. & Rasmussen, T. (1952). *The cerebral cortex of man.* Macmillan, New York.                    [124]

Penfield, W. see also Foerster, O., Hécaen, H.

Penman, G. G. see Clark, W. E. le Gros.

Penn, R. D. & Hagins, W. A. (1969). Signal transmission along retinal rods and the origin of the electroretinographic a-wave. *Nature*, **223**, 201–205.                    [50, 51, 52]

Peters, A. & Palay, S. L. (1966). The morphology of laminae A and A₁ of the dorsal nucleus of the lateral geniculate body of the cat. *J. Anat.*, **100**, 451–486.                    [100]

Pettigrew, J. D. see Barlow, H. B.

Pickles, E. G. see Hecht, S.

Piercy, M. F. see MacFie, J.

Piéron, H. (1929). Influence de la composition de la lumière sur la nature des couleurs subjectives de Fechner-Benham. *Année psychol.*, **29**, 229–233.                    [154]

Piper, H. (1903). Über die Abhängigkeit des Reizwertes leuchtender Objekte von ihrer Flächen- bezw. Winkelgrösse. *Z. psych. u. Physiol. d. Sinnesorg.*, **32**, 98–112.                    [164]

— (1911). Über die Netzhautströme. *Arch. Anat. Physiol. (Lpz.). Physiol. Abt.*, **1911**, 85–132.                    [61, 71]

Pirenne, M. H. (1948). Independent light-detectors in the peripheral retina. *J. Physiol.*, **107**, 47P.                    [193]

Pirenne, M. H. & Marriott, F. H. C. (1955). Absolute threshold and frequency-of-seeing curves. *J. opt. Soc. Amer.*, **45**, 909–912.    [190]

Pirenne, M. H. see also Denton, E. J., Hecht, S.

Pitt, F. H. G. (1944). The nature of normal trichromatic and dichromatic vision. *Proc. Roy. Soc.*, **132B**, 101–117.                    [225]

Pitt, G. A. J. see Morton, R. A.

Platt, J. R. see Cone, R. A.

Poincelot, R. P., Millar, P. G., Kimbel, R. L. & Abrahamson, E. W. (1969). Lipid to protein chromophore transfer in the photolysis of visual pigments. *Nature*, **221**, 256–257.                    [20, 21]

Poincelot, R. P. & Zull, J. E. (1969). Phospholipid composition and extractability of light- and dark-adapted bovine retinal rod outer segments. *Vision Res.*, **9**, 647–651.                    [4]

Polyak, S. L. (1933). A contribution to the cerebral representation of the retina. *J. comp. Neurol.*, **57**, 541–617.                    [98, 99, 100]

— (1941). *The Retina.* University of Chicago Press.    [105, 168, 258]

Polyak, S. L. and Hayashi, R. (1936). The cerebral representation of the retina in the chimpanzee. *Brain*, **59**, 51–60.                    [98]

Powell, T. P. S. see Brooke, R. N. L., Cowan, W. M.

Pratt, D. see Grellman, K. -H.

Pribram, H. B. & Barry, J. (1956). Further behavioral analysis of parieto-temporo-preoccipital cortex. *J. Neurophysiol.*, **19**, 99–106.    [129]

Pribram, K. H. see Mishkin, M.

Priestley, J. (1772). *The history and present state of discoveries relating to vision, light and colours.* 2 vols. Johnson, London.    [201]

Pritchard, R. see Burns, B. D.

Purkinje, Johann (1823). *Beobachtungen und Versuche zur Physiologie der Sinne.* Erstes Bändchen. J. G. Calve, Prague.    [133, 142, 157]

— (1825). *Beobachtungen und Versuche zur Physiologie der Sinne.* Zweites Bändchen. G. Reimer, Berlin.    [133, 136, 159]

Quilliam, J. P. see Lythgoe, R. J.

Rabedeau, R. G. see Cornsweet, T. N.

Raab, D. H. see Ades, H. W.

Radding, C. M. & Wald, G. (1958). The action of enzymes on rhodopsin. *J. gen. Physiol.*, **42**, 371–383.    [4]

Ramón y Cajal, S. see Cajal, S. Ramón.

Ranson, S. W. & Magoun, H. W. (1933). The central path of the pupillo-constrictor reflex in response to light. *Arch. Neurol. Psychiat.*, **30**, 1193–1204.    [94]

Ratliff, F. (1965). *Mach bands: quantitative studies on neural networks in the retina.* San Francisco, Holden-Day.    [153]

Ratliff, F. see Riggs, L. A.

Riccó, A. (1877). Relazione fra il minimo angolo visuale e l'intensitá luminosa. *Annali di Ottalmologia*, **6**, 373–479.    [163]

Richardson, T. M. (1969). Cytoplasmic and ciliary connections between the inner and outer segments of mammalian visual receptors. *Vision Res.*, **9**, 727–731.    [24]

Richardson-Robinson, F. (1923). A case of colour-blindness to yellow and to blue. *Amer. J. Psychol.*, **34**, 157–184.    [226]

Riddoch, G. (1935). Visual disorientation in homonymous half-fields. *Brain*, **58**, 376–382.    [131]

Riggs, L. A., Berry, R. N. & Wayner, M. (1949). A comparison of electrical and psychophysical determinations of the spectral sensitivity of the human eye. *J. opt. Soc. Amer.*, **39**, 427.    [73]

Riggs, L. A. & Johnson, E. P. (1949). Electrical responses of the human retina. *J. exp. Psychol.*, **39**, 415–424.    [71]

Riggs, L. A., Ratliff, F., Cornsweet, J. C. & Cornsweet, T. N. (1953). The disappearance of steadily fixated visual test objects. *J. opt. Soc. Amer.*, **43**, 495–501.    [148]

Riggs, L. A. see also Armington, J. C., Boynton, R. M.

Ripps, H. & Weale, R. A. (1969). Flash bleaching of rhodopsin in the human retina. *J. Physiol.*, **200**, 151–159.    [31]

Ripps, H. see also Ikeda, H.

Rochon-Duvigneaud, A. (1943). *Les yeux et la vision des vertébrés.* Masson, Paris.    [258]

Rodger, F. C. see Hallett, P. E.

Rodieck, R. W. (1967). Receptive fields in the cat retina: a new type. *Science*, **157**, 90–92.                                                    [89]

Rodieck, R. W. see also Kozak, W.

Rodriguez-Peralta, L. A. (1968). Hematic and fluid barriers of the retina and vitreous body. *J. comp. Neurol.*, **132**, 109–124.          [66]

Rose, A. (1948). The sensitivity performance of the human eye on an absolute scale. *J. opt. Soc. Amer.*, **38**, 196–208.          [168, 196]

Rosvold, H. E. see Ettlinger, G.

Rushton, D. N. see Brindley, G. S.

Rushton, W. A. H. (1949). The structure responsible for action potential spikes in the cat's retina. *Nature*, **164**, 743–744.          [82]

— (1955). Foveal photopigments in normal and colour-blind. *J. Physiol.*, **129**, 41P–42P.          [39, 40]

— (1956*a*). The difference spectrum and the photosensitivity of rhodopsin in the living human eye. *J. Physiol.*, **134**, 11–29.          [29, 258]

— (1956*b*). The rhodopsin density in the human rods. *J. Physiol.*, **134**, 30–46.          [162]

— (1957). Blue light and the regeneration of human rhodopsin *in situ. J. gen. Physiol.*, **41**, 419–428.          [34]

— (1961). Rhodopsin measurement and dark-adaptation in a subject deficient in cone vision. *J. Physiol.*, **156**, 193–205.          [180, 181]

— (1963*a*). A cone pigment in the protanope. *J. Physiol.*, **168**, 345–359.          [40, 42, 228]

— (1963*b*). The density of chlorolabe in the foveal cones of the protanope. *J. Physiol.*, **168**, 360–373.          [40, 254, 258]

— (1963*c*). Cone pigment kinetics in the protanope. *J. Physiol.*, **168**, 374–388.          [31, 40, 251]

— (1963*d*). Dark adaptation after exposing the eye to an instantaneous flash. *Nature*, **199**, 971–972.          [184]

— (1965*a*). A foveal pigment in the deuteranope. *J. Physiol.*, **176**, 24–37.          [40, 42, 228]

— (1965*b*). Cone pigment kinetics in the deuteranope. *J. Physiol.*, **176**, 38–45.          [40, 43, 251]

— (1965*c*). Bleached rhodopsin and visual adaptation. *J. Physiol.*, **181**, 645–655.          [182, 183, 184]

— (1968). Rod/cone rivalry in pigment regeneration. *J. Physiol.*, **198**, 219–236.          [44]

Rushton, W. A. H. & Baker, H. D. (1963). Effect of a very bright flash on cone vision and cone pigments in man. *Nature*, **200**, 421–423.          [31, 180, 184]

Rushton, W. A. H., Campbell, F. W., Hagins, W. A. & Brindley, G. S. (1955). The bleaching and regeneration of rhodopsin in the living eye of the albino rabbit and of man. *Optica Acta*, **1**, 182–190.          [28, 34]

Rushton, W. A. H. & Henry, G. H. (1968). Bleaching and regeneration of cone pigments in man. *Vision Res.*, **8**, 617–631.          [28]

Rushton, W. A. H. & Westheimer, G. (1962). The effect upon the rod

threshold of bleaching neighbouring rods. *J. Physiol.*, **164**, 318–329.
[181]

Rushton, W. A. H. see also Baker, H. D., Blakemore, C. B., Brindley, G. S., Campbell, F. W., Donner, K. O., Naka, K. J.

Rozzi, V. P. see Pak, W. L.

Sachs, M. (1891). Über die specifische Lichtabsorbtion des gelben Fleckes der Netzhaut. *Pflügers Arch. ges. Physiol.*, **50**, 574–586. [140]

Sakakura, H. & Iwama, K. (1965). Presynaptic inhibition and post-synaptic facilitation in lateral geniculate body and so-called deep sleep wave activity. *Tohoku J. exp. Med.*, **87**, 40–51.          [103]

Sakakura, H. see also Iwama, K.

Saur, G. see Grüsser, O.-J.

Scherrer, J. see Calvet, J.

Schilder, P. see Pasik, P.

Schmidt, W. J. (1938). Polarisationsoptische Analyse eines Eiweiss-Lipoid-Systems, erläutert am Aussenglied der Sehzellen. *Kolloidzschr.*, **85**, 137–148.          [23]

Schneider, E., Goodeve, C. F. & Lythgoe, R. J. (1939). The spectral variation of the photosensitivity of visual purple. *Proc. Roy. Soc.*, **170A**, 102–112.          [12]

Schneider, G. E. (1969). Two visual systems. *Science*, **163**, 895–902.
[94]

Schön, W. (1874). Einfluss der Ermüdung auf die Farbenempfindung. *v. Graefes Arch. Ophthal.*, **20**, *Abt.* **2**, 273–286.          [151]

Schoolman, A. & Evarts, E. V. (1959). Responses to lateral geniculate radiation stimulation in cats with implanted electrodes. *J. Neurophysiol.*, **22**, 112–129.          [112]

Schrader, M. E. G. (1889). Zur Physiologie des Vogelgehirns. *Pflügers Arch. ges. Physiol.*, **44**, 175–238.          [92]

Schultze, M. (1866). Zur Anatomie und Physiologie der Retina. *Arch. mikr. Anat.*, **2**, 175–286.          [39]

Schum, H. see Krause, F.

Schutta, H. S. see Cole, M.

Sefton, A. J. (1968). The innervation of the lateral geniculate nucleus and anterior colliculus in the rat. *Vision Res.*, **8**, 867–881.          [91]

Sefton, A. J. (*née* Jervie) see also Burke, W., Hayhow, W. R.

Ségal, J. (1953). *Le mécanisme de la vision des couleurs.* G. Doin, Paris.
[139]

Sekuler, R. W. see Pantle, A. J.

Seneviratne, K. N. & Whitteridge, D. (1962). Visual evoked responses in the lateral geniculate nucleus. *Electroenceph. clin. Neurophysiol.*, **14**, 785.          [100]

Seneviratne, K. N. see also Bilge, M.

Shaw, T. I. see Crescitelli, F.

Sherrington, C. S. (1904). On binocular flicker and the correlation of activity of 'corresponding' retinal points. *Brit. J. Psychol.*, **1**, 26–60.          [174]

Shichi, H., Lewis, M. S., Irreverre, F. & Stone, A. L. (1969). Biochemistry of visual pigments: I Purification and properties of bovine rhodopsin. *J. biol. Chem.*, **244**, 529–536. [4, 5]

Shlaer, S., Smith, E. L. & Chase, A. M. (1942). Visual acuity and illumination in different spectral regions. *J. gen. Physiol.*, **25**, 553–569. [167]

Shlaer, S. see also Hecht, S., Lamar, E. S.

Short, A. D. (1966). Decremental and incremental visual thresholds. *J. Physiol.*, **185**, 646–654. [176]

Shukolyukov, S. A. see Etingov, R. N.

Shurcliff, W. A. (1955). Haidinger's brushes and circularly polarised light. *J. opt. Soc. Amer.*, **45**, 399. [141]

Sickel, W. (1966). pp. 115–124 of 'Clinical Electroretinography', Suppl. to *Vision Research* (Pergamon Press). [72]

Sidman, R. L. (1957). The structure and concentration of solids in photoreceptor cells studied by refractometry and interference microscopy. *J. Biophys. biochem. Cytol.*, **3**, 15–30. [25]

Siegel, I. M. see Arden, G. B.

Sjöstrand, F. S. (1953). The ultrastructure of the outer segments of rods and cones of the eye as revealed by the electron microscope. *J. cell. comp. Physiol.*, **42**, 15–44. [23]

Sklyarevich, V. V. see Fedorov, N. T.

Skrebitskiĭ, V. G. see Kuman, E. A.

Sloan, L. L. (1968). The photopic acuity-luminance function with special reference to parafoveal vision. *Vision Res.*, **8**, 901–911. [166]

Smirnov, M. S. see Bongard, M. M.

Smith, E. L. see Chase, A. M., Shlaer, S.

Smith, K. U. & Bridgeman, M. (1943). The neural mechanisms of movement vision and optic nystagmus. *J. exp. Psychol.*, **33**, 165–187. [93]

Smith, P. H. see Wald, G.

Smythies, J. R. (1959). The stroboscopic patterns. *Brit. J. Psychol.*, **50**, 106–116. [157]

Sparrock, J. M. B. see Barlow, H. B.

Sperling, G. (1960). Negative afterimage without prior positive image. *Science*, **131**, 1613–1614. [149]

Sperling, H. G. & Joliffe, C. L. (1965). Intensity-time relationship at threshold for spectral stimuli in human vision. *J. opt. Soc. Amer.*, **55**, 191–199. [169]

Sperry, R. W., Miner, Nancy & Myers, R. E. (1955). Visual pattern perception following subpial slicing and tantalum wire implantations in the visual cortex. *J. comp. physiol. Psychol.*, **48**, 50–58. [125]

Spinelli, D. N., Pribram, K. H. & Weingarten, M. (1965). Centrifugal optic nerve responses evoked by auditory and somatic stimulation. *Exp. Neurol.*, **12**, 303–319. [108, 109]

Spivey, B. E. see Alpern, M.

Sprague, J. M. & Meikle, T. H. (1965). The role of the superior colliculus in visually guided behavior. *Exp. Neurol.*, **11**, 115–146. [93]

Steiner, J. see Kühne, W.

Sterling, P. see Wickelgren, B. G.

Stevens, S. S. (1951). *Handbook of experimental psychology*. Chapman & Hall, London.                                                      [134]

St George, R. C. see Hubbard, R., Wald, G.

Stiles, W. S. (1939). The directional sensitivity of the retina and the spectral sensitivities of the rods and cones. *Proc. Roy. Soc.*, **127B**, 64–105.                                    [162, 187, 188, 231, 248, 249]

— (1944). Current problems of visual research. *Proc. phys. Soc.*, **56**, 329–356.                                                              [210]

— (1946). A modified Helmholtz line-element in brightness-colour space. *Proc. phys. Soc.*, **58**, 41–65.                                    [232]

— (1949). Increment thresholds and the mechanisms of colour vision. *Docum. ophthal.*, **3**, 138–163.        [167, 231, 232, 233, 241, 244]

— (1953). Further studies of visual mechanisms by the two-colour threshold technique. Pp. 65–103 of *Coloquio sobre problemas opticas de la visión*. Union internationale de physique pure et appliquée, Madrid.                                      [140, 231, 234, 235, 236, 256]

— (1955). The basic data of colour matching. Pp 44–65 of *Physical Society Yearbook*, **1955**. Physical Society, London.   [141, 203, 204, 215, 258]

— (1959). Colour vision: the approach through increment-threshold sensitivity. *Proc. Nat. Acad. Sci.*, **45**, 100–114.               [231]

Stiles, W. S. & Burch, J. M. (1959). N.P.L. colour-matching investigations: final report. *Optica Acta*, **6**, 1–26.                     [203]

Stiles, W. S. & Crawford, B. H. (1933). The luminous efficiency of rays entering the eye pupil at different points. *Proc. Roy. Soc.*, **112B**, 428–450.                                                         [248]

Stiles, W. S. see also Aguilar, M., Boynton, R. M., Enoch, J. M., Flamant, F. and p. 134.

Stone, A. L. see Shichi, H.

Stone, J. (1966). The naso-temporal division of the cat's retina. *J. comp. Neurol.*, **126**, 585–600.                                        [91, 127]

Stone, J. & Fabian, M. (1966). Specialized receptive fields of the cat's retina. *Science*, **152**, 1277–1279.                            [88]

Stone, J. & Hansen, S. M. (1966). The projection of the cat's retina on the lateral geniculate nucleus. *J. comp. Neurol.*, **126**, 601–624.   [100]

Stone, J. see also Leicester, J.

Straschill, M. & Taghavy, A. (1967). Neuronale Reaktionen im Tectum opticum der Katze auf bewegte und stationäre Lichtreize. *Exp. Brain Res.*, **3**, 353–367.                                        [94]

Strata, P. see Angel, A.

Sulzer, D. see Broca, A.

Sunga, R. see Enoch, J.

Suzuki, H. & Kato, E. (1965). Cortically induced presynaptic inhibition in cat's lateral geniculate body. *Tohoku J. exp. Med.*, **86**, 277–289.                                                          [102]

Suzuki, H. & Taira, N. (1961). Effect of reticular stimulation upon synpatic transmission in the cat's lateral geniculate body. *Jap. J. Physiol.*, **11**, 641–655.                                         [95]

Svaetichin, G. (1953). The cone action potential. *Acta physiol. Scand.*, **29**, Suppl. **106**, 565–600. [47, 76]
— (1956). Spectral response curves from single cones. *Acta physiol. Scand.*, 39, Suppl. **134**, 17–46. [47]
— see also Granit, R., MacNichol, E. J.
Symmes, D. see Anderson, K. V.
Szentágothai, J., Hámori, J. & Tömböl, T. (1966). Degeneration and electron microscope analysis of the synaptic glomeruli in the lateral geniculate body. *Exp. Brain Res.*, **2**, 283–301. [95, 98, 100]
Taghavy, A. see Straschill, M.
Takagi, M. (1963). Studies on the ultraviolet spectral displacements of cattle rhodopsin. *Biochim. Biophys. Acta.*, **66**, 328–340. [5]
Takahashi, E. see Levick, W. R.
Taira, N. see Suzuki, H.
Talbot, S. A. (1942). A lateral localization in cats' visual cortex. *Fed. Proc.*, **1**, 84. [111]
Talbot, S. A. & Marshall, W. H. (1941). Physiological studies on neural mechanisms of visual localization and discrimination. *Amer. J. Ophthal.*, **24**, 1255–1263. [110]
Talbot, S. A. see also Marshall, W. H.
Tansley, K. (1931). The regeneration of visual purple: its relation to dark adaptation and night blindness. *J. Physiol.*, **71**, 442–458. [34]
Tansley, K. see Parry, H. B.
Tasaki, K. (1960). Some observations on the retinal potentials of the fish. *Arch. ital. Biol.*, **98**, 81–91. [78]
Teller, D. Y. (1968). Increment threshold on black bars. *Vision Res.*, **8**, 713–718. [185]
Teller, D. Y., Andrews, D. P. & Barlow, H. B. (1966). Local adaptation in stabilized vision. *Vision Res.*, **6**, 701–705. [182, 185]
Tello, F. (1904). Disposición macroscópica y estructura del cuerpo geniculado externo. *Trab. Lab. Invest. biol. Univ. Madrid*, **3**, 39–62. [97, 98]
Templeton, W. B. & Green, F. A. (1968). Chance results in utrocular discrimination. *Q. J. exp. Psy.*, **20**, 200–203. [152, 248]
Teuber, H. -L., Battersby, W. S. & Bender, M. B. (1960). *Visual field defects after penetrating missile wounds of the brain.* Harvard Univ. Press. [110, 174]
Thomas, G. J. (1955). A comparison of uniocular and binocular critical flicker frequencies: simultaneous and alternate flashes. *Amer. J. Psychol.*, **68**, 37–53. [174]
Thomson, L. C. & Wright, W. D. (1947). The colour sensitivity of the retina within the central fovea of man. *J. Physiol.*, **105**, 316–331. [240]
Thomson, L. C. see also Parry, H. B.
Tilly, R. see Pedler, C. M. H.
Toida, N. see Goto, M.
Tömböl, T. see Szentágothai, J.
Tomita, T. (1950). Studies on the intraretinal action potential Part I:

Relation between the localisation of micro-pipette in the retina and the shape of the intraretinal action potential. *Jap. J. Physiol.*, **1**, 110–117. [64, 67, 68]

Tomita, T. (1957). A study on the origin of intraretinal action potential of the cyprinid fish by means of pencil-type microelectrode. *Jap. J. Physiol.*, **7**, 80–85. [47, 76]

— (1963). Electrical activity in the vertebrate retina. *J. opt. Soc. Amer.*, **53**, 49–57. [68]

— (1965). Electrophysiological study of the mechanisms subserving color coding in the fish retina. *Cold. Spr. Harb. Symp.*, **30**, 559–566. [48, 69]

Tomita, T. & Kaneko, A. (1965). An intracellular coaxial microelectrode. Its construction and application. *Med. Electron. biol. Engng.*, **3**, 367–376. [78]

Tomita, T., Kaneko, A., Murakami, M. & Pautler, E. L. (1967). Spectral response curves of single cones in the carp. *Vision Res.*, **7**, 519–531. [48, 53]

Tomita, T. see also Hashimoto, Y., Toyoda, J.

Toraldo di Francia, G. (1948). Per una teoria dell'effetto Stiles-Crawford. *Nuovo Cimento*, Series 9, **5**, 589–590. [252]

Toraldo di Francia, G. see also Fiorentini, A.

Torii, S. see Alpern, M.

Tosaka, T. see Watanabe, K.

Toyoda, J., Nosaki, H. & Tomita, T. (1969). Light-induced resistance changes in single photoreceptors of Necturus and Gekko. *Vision Res.*, **9**, 453–463. [48, 49]

Trezona, P. W. (1953). Additivity of colour equations. *Proc. phys. Soc.*, **66B**, 547–556. [216]

— (1954). Additivity of colour equations: II. *Proc. phys. Soc.*, **67B**, 513–522. [215, 216]

Trifonov, Yu. A. (1968). An investigation of synaptic transmission between the photoreceptor and the horizontal cell with the help of electrical stimulation of the retina (in Russian). *Biofizika*, **13**, 809–816. [80]

Trifonov, Yu. A. & Utina, I. A. (1966). Investigation of the mechanism of action of currents on L-cells of the retina (in Russian). *Biofizika*, **11**, 646–652. [157]

Trifonov, Yu. A. see also Byzov, A. L., Ostrovskiï, M. A.

Troxler, D. (1804). Ueber das Verschwinden gegebener Gegenstände innerhalb unseres Gesichtskreises. *Ophthalm. Bibliothek.* (ed. K. Himly & J. A. Schmidt), **2/2**, 1–53. [148]

Tschermak, A. (1898). Über die Bedeutung der Lichtstärke und des Zustandes des Sehorgans für farblose optische Gleichungen. *Pflügers Arch. ges. Physiol.*, **70**, 297–328. [204, 215, 258]

Utina, I. A. (1960). Investigation of the activity of bipolar cells of two types by ultraviolet cytophotometry (in Russian). *Biofizika*, **5**, 626–627. [60]

— (1964). Changes in the quantity of RNA in horizontal and amacrine cells of the frog's retina under various conditions of illumination (in Russian). *Doklady Akad. Nauk.*, **157**, 1216–1218.    [61]

Utina, I. A. & Byzov, A. L. (1965). Investigation of the functional properties of photoceptors of the frog's retina by a cytochemical method (in Russian). *Biofizika*, **10**, 1088–1091.    [61]

Utina, I. A., Nechaeva, N. V. & Brodskiĭ, V. Ya. (1960). RNA in ganglion cells of the frog's retina in darkness and in steady and flashing light (in Russian). *Biofizika*, **5**, 749–750.    [60]

Utina, I. A. see also Trifonov, Yu. A.

Vakkur, G. J. see Bishop, P. O.

Vastola, E. F. (1961). A direct pathway from lateral geniculate body to association cortex. *J. Neurophysiol.*, **24**, 469–487.    [111]

van der Velden, H. A. (1944). Over het aantal lichtquanta dat nodig is voor een lichtprikkel bij het menselijk oog. *Physica*, **11**, 179–189.    [191, 192]

— see also Bouman, M. A.

Vernon, M. see Craik, K.

Verriest, G. see François, J.

da Vinci, L. see Leonardo da Vinci.

Voeste, H. (1898). Messende Versuche über die Qualitätsänderungen der Spectralfarben in Folge von Ermüdung der Netzhaut. *Z. Psychol. Physiol. Sinnesorg*, **18**, 257–267.    [151]

Vogt, M. see Feldberg, W.

Volkmann, F. C. (1962). Vision during voluntary saccadic eye movements. *J. opt. Soc. Amer.*, **52**, 571–578.    [186]

Wagner, H. G., MacNichol, E. F. & Wolbarsht, M. L. (1960). The response properties of single ganglion cells in the goldfish retina. *J. gen. Physiol.*, **43**, 45–62.    [84]

Wald, G. (1935). Carotenoids and the visual cycle. *J. gen. Physiol.*, **19**, 351–371.    [14, 31]

— (1937). Photo-labile pigments of the chicken retina. *Nature*, **140**, 545–546.    [2, 39]

— (1938). On rhodopsin in solution. *J. gen. Physiol.*, **21**, 795–832.    [3, 18]

— (1939). The porphyropsin visual system. *J. gen. Physiol.*, **22**, 775–794.    [36]

— (1944). Vision: photochemistry. Pp. 1658–1667 of *Medical Physics*, ed. O. Glasser. Year Book Publishers, Chicago.    [180]

— (1945). Human vision and the spectrum. *Science*, **101**, 653–658.    [140, 162]

— (1954). On the mechanism of the visual threshold and visual adaptation. *Science*, **119**, 887–892.    [181]

— (1967). Blue-blindness in the normal fovea. *J. opt. Soc. Amer.*, **57**, 1289–1301.    [241, 244]

— (1968). The molecular basis of visual excitation. *Nature*, **219**, 800–807.    [21]

Wald, G. & Brown, P. K. (1953). The molar extinction of rhodopsin. *J. gen. Physiol.*, **37**, 189–200.    [5, 13]

Wald, G. & Brown, P. K. (1956). Synthesis and bleaching of rhodopsin. *Nature*, **177**, 174–176.      [17]

— (1958). Human rhodopsin. *Science*, **127**, 222–226.      [26, 162]

Wald, G., Brown, P. K. & Kennedy, D. (1957). The visual system of the alligator. *J. gen. Physiol.*, **40**, 703–713.      [180]

Wald, G., Brown, P. K. & Smith, P. H. (1955). Iodopsin. *J. gen. Physiol.*, **38**, 623–681.      [2, 39]

Wald, G., Durell, J. & St George, R. C. C. (1950). The light reaction in the bleaching of rhodopsin. *Science*, **111**, 179–181.      [9, 10, 23]

Wald, G. see also Auerbach, E., Brown, P. K., Hubbard, R., Matthews, R. G., Yoshizawa, T.

Walker, M. A. see Denton, E. J.

Wall, P. D. (1958). Excitability changes in afferent fibre terminations and their relation to slow potentials. *J. Physiol.*, **142**, 1–21.      [102]

Wallach, H. see Kohler, W.

Waller, R. (1686). A catalogue of simple and mixt colours, with a specimen of each colour prefixt to its proper name. *Phil. Trans.*, **16**, 24–32.      [200, 201]

Walley, R. E. (1967). Receptive fields in the accessory optic system of the rabbit. *Exp. Neurol.*, **17**, 27–43.      [95]

Walls, G. L. & Mathews, R. W. (1952). New means of studying color blindness and normal foveal color vision. University of California Publications in Psychology, Volume 6, No. 1.      [140]

Walraven, P. L. (1966). Recovery from the increase of the Stiles-Crawford effect after bleaching. *Nature*, **210**, 311–312.      [249]

Walters, H. V. (1942). Some experiments on the trichromatic theory of vision. *Proc. Roy. Soc.*, **131B**, 27–49.      [225, 245]

Walters, H. V. & Wright, W. D. (1943). The spectral sensitivity of the fovea and extrafovea in the Purkinje range. *Proc. Roy. Soc.*, **131B**, 340–361.      [162]

Warren, F. J. see Denton, E. J.

Warrington, E. K., James, M. & Kinsbourne, M. (1966). Drawing disability in relation to laterality of cerebral lesion. *Brain.* **89**, 53–82.      [131]

Warrington, E. K. see also Cole, M.

Wasserman, G. S. (1966). Brightness enhancement in intermittent light: variation of luminance and light dark ratio. *J. opt. Soc. Amer.*, **56**, 242–250.      [143]

Watanabe, K. & Tosaka, T. (1959). Functional organization of the cyprinid fish retina as revealed by discriminative responses to spectral illumination. *Jap. J. Physiol.*, **9**, 84–93.      [78]

Watanabe, K., Tosaka, T. & Yokota, T. (1960). Effects of extrinsic electric current on the cyprinid fish EIRG (s-potential). *Jap. J. Physiol.*, **9**, 132–141.      [78]

Watanabe, K. see also Brown, K. T.

Wayner, M. see Riggs, L. A.

Weale, R. A. (1953a). Spectral sensitivity and wavelength discrimination of the peripheral retina. *J. Physiol.*, **119**, 170–190.      [163]

Weale, K. A. (1953b). Photochemical reactions in the living cat's retina. *J. Physiol.*, **122**, 322–331. [29]

— (1953c). Cone-monochromatism. *J. Physiol.*, **121**, 548–569. [227]

— (1959). Photosensitive reactions in fovea of normal and cone-monochromatic observers. *Optica Acta.*, **6**, 158–174. [229]

— (1967). On an early stage of rhodopsin regeneration in man. *Vision Res.*, **7**, 819–827. [29]

— see also Arden, G. B., Ripps, H.

Webb, C. see Hayhow, W. R

Weber, E. H. (1834). *De pulsu, resorptione, auditu et tactu annotationes anatomicae et physiologicae.* C. F. Koehler, Leipzig. [175]

Weinstein, G. W., Hobson, R. R. & Dowling, J. E. (1967). Light and dark adaptation in the isolated rat retina. *Nature*, **215**, 134–138. [34]

Weiskrantz, L. & Mishkin, M. (1958). Effects of temporal and frontal cortical lesions on auditory discrimination in monkeys. *Brain*, **81**, 406–414. [129]

Weiskrantz, L. see Cowey, A., Humphrey, N. K.

Werblin, F. S. & Dowling, J. E. (1969). Organization of the retina of the mudpuppy, Necturus maculosus: II. Intra-cellular recording. *J. Neurophysiol.*, **32**, 339–355. [48, 75, 77, 78]

Westheimer, G. (1965). Spatial interaction in the human retina during scotopic vision. *J. Physiol.*, **181**, 881–894. [182]

— (1967a). Spatial interaction in human cone vision. *J. Physiol.*, **190**, 139–154. [185]

— (1967b). Dependence of the magnitude of the Stiles-Crawford effect on retinal location. *J. Physiol.*, **192**, 309–315. [249]

— (1968a). Bleached rhodopsin and retinal interaction. *J. Physiol.*, **195**, 97–105. [182, 183]

— (1968b). Entoptic visualization of Stiles-Crawford effect. *Arch. Ophthal.*, **79**, 584–588. [136]

— see also Brindley, G. S., Rushton, W. A. H.

Whalen, R. E. see Cornsweet, T. N.

Whitteridge, D. see Bilge, M., Choudhury, B. P., Seneviratne, K. N.

Whitty, C. W. M. & Newcombe, F. (1965). Disabilities associated with lesions in the posterior parietal region of the non-dominant hemispheres. *Neuropsychologia*, **3**, 175–185. [131]

Widén, L. & Ajmone Marsan, C. (1960). Unitary analysis of the response elicited in the visual cortex of cat. *Arch. ital. Biol.*, **98**, 248–274. [112]

Wickelgren, B. G. & Sterling, P. (1969). Influence of visual cortex on receptive fields in the superior colliculus of the cat. *J. Neurophysiol.*, **32**, 16–23. [93]

Wiesel, T. N. (1960). Receptive fields of ganglion cells in the cat's retina. *J. Physiol.*, **153**, 583–594. [87]

Wiesel, T. N. & Hubel, D. H. (1966). Spatial and chromatic interactions in the lateral geniculate body of the rhesus monkey. *J. Neurophysiol.*, **29**, 1115–1156. [103, 104]

Wiesel, T. N. see also Brown, K. T., Hubel, D. H.

Wilkins, M. H. F. see Blaurock, A. E.

Williams, D. R. see Cornsweet, T. N.

Willmer, E. N. (1946). *Retinal Structure and Colour Vision*. Cambridge University Press. [257]

— (1950). The monochromatism of the central fovea in red-green-blind subjects. *J. Physiol.*, **110**, 377–385. [244]

— (1954). Subjective brightness and size of field in the central fovea. *J. Physiol.*, **123**, 315–323. [143]

Willmer, E. N. & Wright, W. D. (1945). Colour sensitivity of the fovea centralis. *Nature, Lond.*, **156**, 119–121. [240]

Willmer, E. N. see also Brindley, G. S.

Wilson, M. E. see Choudhury, B. P.

Winans, S. S. (1967). Visual form discrimination after removal of the visual cortex in cats. *Science*, **158**, 944–946. [92, 125]

Witkovsky, P. (1967). A comparison of ganglion cell and S-potential response properties in the carp retina. *J. Neurophysiol.*, **30**, 546–561. [79]

Wohlgemuth, A. (1911). On the after-effect of seen movement. *Brit. J. Psychol. Monogr. Suppl.*, **1**, 1–117. [145]

Wolbarsht, M. L. see Wagner, H. G.

Wolf, E. see Berkley, M.

Wolff, H. G. & Lennox, W. G. (1930). Cerebral circulation. XIII. The effect on pial vessels of variations in the oxygen and carbon dioxide content of the blood. *Arch. Neurol. Psychiat. (Chicago)*, **23**, 1097–1120. [123]

Wolff, J. G., Delacour, J., Carpenter, R. H. S. & Brindley, G. S. (1968). The patterns seen when alternating electric current is passed through the eye. *Q. J. exp. Psychol.*, **20**, 1–10. [158]

Wolter, J. R. & Liss, L. (1956). Zentrifugale (antidrome) Nervenfasern im menschlichen Sehnerven. *v. Graefes Arch. Ophthal.*, **158**, 1–7. [106]

Wolter, J. R. see also Falls, H. F.

Wright, W. D. (1929). A re-determination of the trichromatic mixture data. *Med. Res. Counc. Spec. Rep.*, No. *139*, pp 1–38. [203]

— (1934). The measurement and analysis of colour adaptation phenomena. *Proc. Roy. Soc.*, **115B**, 49–87 [151, 223, 245]

— (1936). The breakdown of a colour match with high intensities of adaptation. *J. Physiol.*, **87**, 23–33. [216]

— (1946). *Researches on Normal and Defective Colour Vision*. Henry Kimpton, London. [204, 223, 224, 227]

— (1952). The characteristics of tritanopia. *J. opt. Soc. Amer.*, **42**, 509–521. [228, 237]

Wright, W. D. & Nelson, J. H. (1936). The relation between the apparent intensity of a beam of light and the angle at which the beam strikes the retina. *Proc. phys. Soc. Lond.*, **48**, 401–405. [252]

Wright, W. D. see also Thomson, L. D., Walters, H. V., Willmer, E. N.

Wyllie, J. H. see Denton, E. J.

Yokota, T. see Watanabe, K.

Yoshizawa, T. & Wald, G. (1963). Pre-lumirhodopsin and the bleaching of visual pigments. *Nature*, **197**, 1279–1286. [12]

— (1967). Photochemistry of iodopsin. *Nature*, **214**, 566–591. [40]

Yoshizawa, T. see also Hubbard, R.

Young, Thomas (1820*a*). On the theory of light and colours. *Phil. Trans.* **1802**, 12–48. [202]

— (1802*b*). An account of some cases of the production of colours, not hitherto described. *Phil. Trans.*, **1802**, 387–397. [202]

Yuryev, M. A. see Fedorov, N. T.


Zeeman, W. P. C. see Brouwer, B.

Zuckerman, S. see Clark, W. E. Le Gros.

Zull, J. E. see Poincelot, R. P.

Zumft, J. see König, A.


Østerberg, G. indexed as if spelled 'Osterberg'.

# INDEX OF SUBJECTS